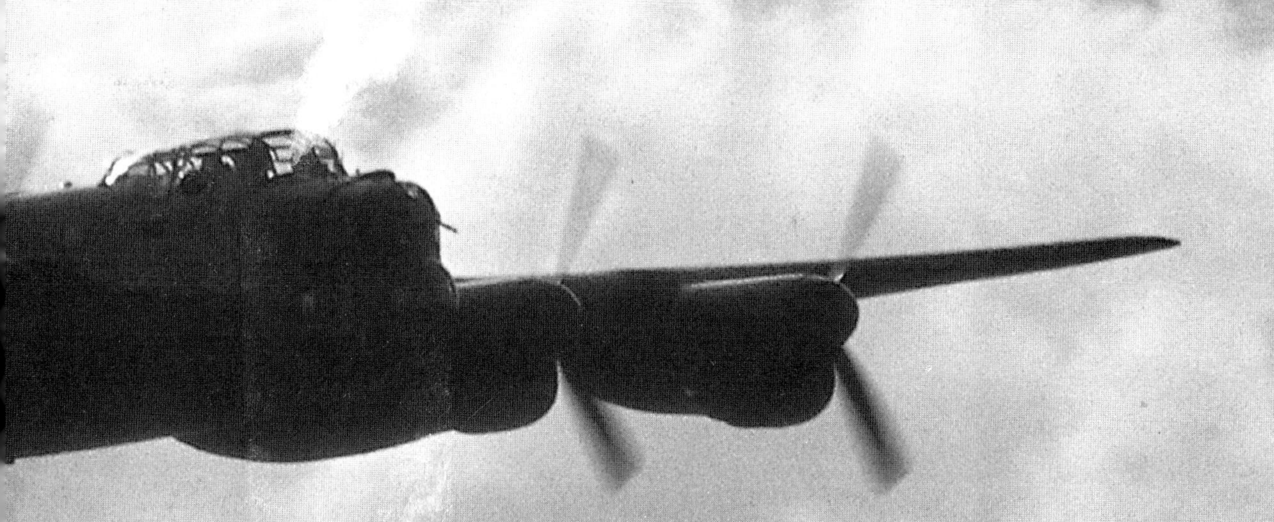

THE ROYAL AIR FORCE AT WAR

Other books by the same author:

Fields of Little America
Encyclopedia of American Military Aircraft
The B-24 Liberator, 1939–45
Castles in the Air
Home by Christmas?
The Bedford Triangle
Wellington: The Geodetic Giant
The World's Fastest Aircraft
Famous Bombers
Classic Fighter Aircraft
Modern Military Aircraft
Flying to Glory
Spirits in the Sky (with Patrick Bunce)
Four Miles High
Great American Air Battles
Thunder in the Heavens (with Patrick Bunce)
8th Air Force at War
The Men Who Flew the Mosquito
Low Level from Swanton
The USAF at War
Confounding the Reich (with Tom Cushing)
USAF Handbook 1939–1945
Raiders of the Reich (with Theo Boiten)

As part of our ongoing market research, we are always pleased to receive comments about our books, suggestions for new titles, or requests for catalogues. Please write to: The Editorial Director, Patrick Stephens Limited, Sparkford, Nr Yeovil, Somerset BA22 7JJ.

THE ROYAL AIR FORCE AT WAR

Memories and personal experiences, 1939 to the present day

MARTIN W. BOWMAN

Patrick Stephens Limited

The brave are surely those who have the clearest vision of what is before them, glory and danger alike, and yet they go out to meet it.

They faced the foe in the strength of their own manhood; and when the shock of battle came they chose rather to suffer the uttermost than to save their lives by feeble submission.

For the whole earth is the sepulchre of heroes; and their story is not only engraved on stone or brass in the places from which they came, their story lives on in places far away, without visible symbol woven into the stuff of other men's lives.

Take these men as your example. Like them, remember that posterity can only be for the free; and that freedom is the sure possession only of those who have the courage to defend it.

Extract from the funeral oration of Pericles, made in 431 BC in memory of Athenians.

First published in March 1997

British Library cataloguing-in-publication data:
A catalogue record of this book is available from the British Library

ISBN 1 85260 540 5

Library of Congress catalog card no. 96-78557

Patrick Stephens Limited is an imprint of Haynes Publishing, Sparkford, Nr Yeovil, Somerset, BA22 7JJ

Designed & typeset by G&M, Raunds, Northamptonshire
Printed and bound in Great Britain by Butler & Tanner Ltd, London and Frome

CONTENTS

CAST OF CHARACTERS

Aaron, Flt Sgt Arthur Louis, VC, DFM, pilot
Allam, Sqn Ldr C.M., Jaguar pilot 41 Squadron
Allen, Denis, 40 Squadron
Anderson, Flt Sgt, pilot, 617 Squadron
Astell, Flt Lt David, DFC, pilot, 617 Squadron
Aubrey-Cound, Flt Lt Rupert, pilot, 209 Squadron
Ayliffe, Sgt Plt E.W.
Bader, Sqn Ldr Douglas, 242 Squadron
Bagshaw, Sqn Ldr Dave, AFC, pilot, 41 Squadron
Bailey, Edward, WOP-AG, 120 Squadron
Ballance, Paul, No. 1 BFTS
Barlow, Flt Lt RAAF, pilot, 617 Squadron
Barton, Plt Off Cyril Joe, RAFVR, pilot
Bateman, Plt Off John, Boston gunner
Bazalgette, Sqn Ldr Ian W., DFC, RAFVR 'master bomber'
Beaudet, Paul, navigator 23 Squadron
Beedie, Wg Cdr Al, OC 55 Squadron
Bennett, AVM D.C.T., CB, CBE, DSO
Best, Wg Cdr R.E., AFC, RAF.
Betts, Gp Capt E.B.C.
Booth, Charles, 38 Squadron Wellington r/o
Bottomley, Air Cdre Norman
Bradfield, Len, 49 Squadron Lancaster air bomber
Branagan, John, WOP-AG, 59 Squadron
Brettell, Flt Lt Edward Gordon
Brettell, Terence
Broadbent, Wg Cdr John, OC Tornado Det., Muharraq
Broom, Sir Ivor, KCB, CBE, DSO, DFC, AFC
Brown, Sqn Ldr W.N., DFC, 12 Squadron
Brown, Flt Sgt Ken, RCAF, pilot, 617 Squadron
Brown, Sqn Ldr A.
Buckle, Herbert, glider pilot and 4 BFTS cadet
Buckley, Gp Capt Nick
Burgess, Flt Sgt Stanley, fitter, 159 Squadron
Burpee, Plt Off, RCAF, pilot, 617 Squadron
Bushell, Sqn Ldr Roger, SAAF 'Big X'
Byers, Flt Sgt, RAAF, pilot, 617 Squadron
Campbell, Fg Off Kenneth, pilot
Carder, Flt Lt Chris, 54 Squadron Jaguar pilot
Cheshire, Wg Cdr Geoffrey L., DSO, DFC, RAFVR, pilot
Churchill, Winston S.
Clark, John, navigator, 192 Squadron
Clarke, AC1 Len R., WOP-AG, 12 Squadron
Cliffe, Wg Cdr J.A., pilot, 11 Squadron
Cooper, LAC W.G. 'Bill', electrician II
Craske, Basil S., RAFVR, pilot, 10 Squadron
Cruickshank, Fg Off John Alexander, pilot
Davis, Dennis 'Joe', electrician, 515 Squadron
Deboni, Victor, No. 1 BFTS
Edwards, Acting Wg Cdr Hughie Idwal, DFC
Ellicott, Fg Off R.A.C., 214 (Valiant) Squadron
Everson, Reg, Arnold Scheme and Mosquito pilot
Fereday, Deryck, pilot, 178 Squadron
Freeman, L. James, cadet pilot, No 1 BFTS
Frost Flt Lt A.S., BSc, navigator, 617 Squadron
Fusniak, Joe, Wellington gunner.
Galland, Adolf, Generalmajor
Gaul, Sgt Wally, Wellington gunner
Gibson, Wg Cdr Guy, VC, DSO, DFC, pilot 617 Squadron
Goebbels, Hermann, Nazi Propaganda Minister
Goering, Hermann
Goldberg, Sgt Harris B., USA, air gunner, RCAF
Gray, Lt Robert Hampton, DSC, RCNVR

Hammersley, Sgt Roland A., DFM, 57 Squadron WOP-AG
Hannah, Flt Sgt John, WOP-AG
Hardeman, John, rear-gunner, 159 Squadron
Harris, ACM Sir A.T., KCB, OBE, AFC, C-in-C Bomber Command
Harrison, Graham, gunner
Hitler, Adolph
Holmes, Pat
Hopgood, Flt Lt J.V. 'Hoppy', pilot, 617 Squadron
Hornell, Flt Lt David Ernest, RCAF
Howland, M., RAAF, pilot 684 Squadron
Irving, Gp Capt N.R., AFC
Jackson, Sgt (later W/O) Norman Cyril, RAFVR
Jenner, Sgt Alfred, WOP-AG, 99 Squadron
Jones, Flt Lt Frank, pilot, Lincoln Flight
Kerss, Wg Cdr Tim J., MBE BSc, OC 54(F) Squadron
Knight, Flt Lt L.E.S., RAAF, pilot, 617 Squadron
Knowles, Flt Lt D.J., BA
Learoyd, Act Flt Lt Roderick Alastair Brook, pilot.
Nicolson, Flt Lt Eric James Brindley
Light, Daphne, WAAF
Livingstone-Spence, David, No. 4 BFTS
Lloyd, AVM Sir Hugh Pughe
Lord, Flt Lt David Samuel Anthony, DFC, pilot
Lub, Jaap, gunner, 320 (Dutch) Squadron
Malcolm, Acting Wg Cdr Hugh Gordon
Maltby, Flt Lt David J.H., DFC, pilot, 617 Squadron
Manser, Fg Off Leslie Thomas, RAFVR, pilot
Mark, Flt Lt John
Martin, Flt Lt Mick, DFC, RAAF, pilot, 617 Squadron
Masters, Sgt Pilot Eric, pilot, 99 Squadron
Maudslay, Sqn Ldr Henry E., pilot, 617 Squadron
McCann, Sgt Ignatius J., RCAF
McCarthy, Flt Lt Joe, USA, pilot, 617 Squadron
McCorkle, Jim, pilot
McDonnell, Gp Capt D.K.L., OBE, RAF
Melling, Larry, pilot, 51 Squadron
Middleton, Flt Sgt Rawdon Hume, RAAF, pilot
Mortimor, Frank, air gunner, 34 (SAAF) Squadron
Munro, Flt Lt Les, RNZAF, pilot, 617 Squadron
Myles, Flt Lt John R., RCAF, 544 Squadron
Mynarski, Plt Off Andrew C., RCAF, mid-upper gunner
Nettleton, Act Sqn Ldr John Deering, pilot
Newton, Flt Lt William Ellis, RAAF
Orford, Cpl A.E.
Ottley, Plt Off, pilot, 617 Squadron
Palmer, Sqn Ldr Robert Anthony M., DFC RAFVR, pilot
Parnell, Geoff, air gunner
Patterson, Boston, RCAF
Petts, Sgt Frank, Wellington pilot, 9 Squadron
Phantom Observer No. 7 Squadron
Price, Cpl Andy, 55 Squadron
Prutton, Sgt Don, B-24 Flt Engineer, 214 Squadron
Reid, Act Flt Lt William, RAFVR, pilot
Rice, Plt Off Allan, pilot, 617 Squadron
Rimer, Fg Off Willie P., Mosquito pilot
Rondot, Sqn Ldr Mike, Jaguar pilot
Scarf, Sqn Ldr Arthur Stewart King, pilot
Searby, Air Cdre John, DSO, DFC
Shannon, Flt Lt Dave J., DFC, RAAF, pilot, 617 Squadron
Short, Gp Cdr Jack A.V., air gunner
Sleep, A. Burford
Smith, Daphne, WAAF
Smith, Derek, Mosquito navigator
Soffe, Flt Lt C.R., BSc, 1(F) Squadron
Stewart, George, RCAF, 23 Squadron Mosquito pilot
Swales, Capt Edwin, DFC, SAAF, 'master bomber'
Tagg, Sgt Harry, 1655 Mosquito Training Unit
Tait, Flt Lt A.G., navigator, XI Squadron
Taylor, Geoff, RAAF, pilot
Thompson, Flt Sgt George, RAFVR, WOP
Thorogood, Sgt Fred, WOP/AG
Townsend, Flt Sgt W.C., pilot, 617 Squadron
Trenchard, Lord
Trent, Sqn Ldr Leonard Henry, DFC, RNZAF, pilot
Trigg, Fg Off Lloyd Alan, DFC, RNZAF, pilot
Trubshaw, Brian, CBE MVO, 4 BFTS
van den Born, Christina
Van der Veen, Gp Capt Marten
Wallace, Sqn Ldr Barry, 120 Squadron
Ward, Sgt James Allen, RNZAF, 2nd pilot
Waterman, Derek, DFC, pilot, 158 Squadron
Williams, Wg Cdr Dil
Williams, Wg Cdr David, CO, 55 Squadron
Williamson, Flt Sgt Ivan, DFC, RNZAF, WOP, 115 Squadron
Wilson, AM Sir Andrew, KCB, AFC, RAF, C-in-C RAF
Wingham, Fred, pilot, 420 Squadron,
Witts, Wg Cdr Jerry J., DSO, MBIM, 31 Squadron, OC Tornados, Dahran
Wood, Plt Off J. Ralph, RCAF, navigator

Woodhouse, David, 70th Suffolk Yeomanry
Woodman, Flt Lt R.G. 'Tim', pilot, 169 Squadron
Young, Sqn Ldr Melvyn 'Dinghy', DFC, pilot, 617 Squadron

SUPPORTING CAST

Allgemeine Zeitung
Adams, Steve
Anderson, R.
Bailey, Mike
Bates, Barbara
Brewell, Sgt Rick
Bulpett Sqn Ldr Ed (RAF Ret'd)
Claxton, Johnny
Collis, Bob
Coward, Noél
Cushing, Tom
Crosby, Frank
Frick, Caroline
Grealy, Audrey
Harvey, John
Hobson, Chris, Snr Librarian, RAF Staff College, Bracknell
Holmes, Harry
Hull, Ralph
Jefferson, Steve
Kinsey, Gordon
Kipling, Rudyard
McCash, Capt Bill, AFM, Falcon Field Assoc
McLoughlin, Sqn Ldr J.E., MBE, BEM, RAF (Ret'd), RAF PR
Mearns, Hughes
Miles, Jasper
Milne, Ian, Wg Cmdr OC 6 Squadron
Milroy Sqn Ldr Hugh, RAF
Norfolk & Norwich Aviation Museum
Ogley, Bob
Percival-Barker, Sqn Ldr Keith
Pile, Stephen
Pope, C/Tech Steve
Pymm, Brian
Russell, Wg Cdr Philip
Thorogood, Ralf
Volkischer Beobachter
Walsh, Colin
Westmacott, D.
White, Ray (Bob)

EXTRAS

Heath Robinson, William, Professor
Kite, Fg Off
KOZ
MAC
Prune, Plt Off
Punch
Tee Emm

THE THIN BLUE LINE

*'Who'll fly a Wimpy, who'll fly a Wimpy,
Who'll fly a Wimpy over Germany?
I, said the Pilot, I said the Pilot,
I'll fly a Hercules mark Three.'*

In September 1939 RAF Bomber Command had fewer than 400 operational bombers, including many obsolescent Handley Page Hampdens and Fairey Battles. It fell to the twin-engined Vickers-Armstrongs Wellingtons and Armstrong Whitworth Whitleys to carry the brunt of the offensive to Germany for the first two years of the war.

'There was obviously something wrong the way this war was being fought and as far as we were concerned, we were being used in a role for which we had never been trained . . .

'In 1939 the officially accepted theory was that fighters had such a small speed advantage over the "modern" bomber that any attack must become a stern chase. It was also accepted that fighters attacking a section of three bombers flying in "Vic" would attack in "Vic" formation . . .'
Sgt Frank Petts, Wellington pilot, 9 Squadron, September 1939.

Plans were laid for the first RAF raid of the war to take place during the afternoon of 4 September 1939. While 15 unescorted Blenheims took off for a strike on the *Admiral von Scheer* at Wilhelmshaven, eight Wellingtons of 9 Squadron and six Wellingtons of 149 Squadron, also without escort, flew on over the North Sea towards Brunsbüttel. Their targets were the battleships *Scharnhorst* and *Gneisenau*, which had earlier been spotted by a Blenheim reconnaissance aircraft from RAF Wyton. Two Wellingtons and five Blenheims failed to return. Worse was to follow.

'It is now by no means certain that enemy fighters did in fact succeed in shooting down any of the Wellingtons . . . the failure of the enemy must be ascribed to good formation flying. The maintenance of tight, unshaken formations in the face of the most powerful

'There was a feeling, voiced mostly by the sergeant captains, that we knew more about the war than either of them." Sergeant Frank Petts, Wellington pilot, right. (Frank Petts)

enemy action is the test of bomber force fighting efficiency and morale. In our Service it is the equivalent of the old "Thin Red Line" or the "Shoulder-to-Shoulder" of Cromwell's

"Following the heavy losses in daylight, Bomber Command Wellingtons, Hampdens and Whitleys were switched to night attacks." (City of Norwich Aviation Museum and BAe)

Ironsides. Had it not been for that good leadership, losses from enemy aircraft might have been heavy.'
Air Cdre Norman Bottomley, Senior Air Staff Officer at Bomber Command HQ, commenting on the loss of six out of 12 Wellingtons of 99 Squadron, 14 December 1939.

'The operation of 18 December, carrying, in search of warships, bombs quite unsuitable for such targets, cost 12 Wellingtons, 11 complete crews and several wounded. Among the 9 Squadron casualties was my Flight Commander and Section Leader, Sqn Ldr Archibald Guthrie. His name is now one of the first on the Air Forces Memorial at Runnymede. Among the surviving crews were Sgt Purdy and his Section Leader Fg Off Grant, and Sgt Ramshawe who had ditched beside a fishing vessel off Hull.

'By the time we returned from ten day's special leave we had a new Flight Commander, Sqn Ldr (later Air Cdre) L.E. Jarman from Training Command. Shortly afterwards a new CO, Wg Cdr McKee (later C-in-C Transport Command) arrived from 99 Squadron. There was a feeling, voiced mostly by the sergeant captains, that we knew more about the war than either of them.'
Sgt Frank Petts, Wellington pilot, 9 Squadron, 1939.

Following the heavy losses in daylight Bomber Command Wellingtons, Hampdens and Whitleys were switched to night attacks, just as the Luftwaffe had done. Bomber Command became largely a night-bombing force, but losses continued to rise.

Our Friend is the Night
Our friend is the night
It hinders their sight
Though their guns still fire.
We never shall tire
In our search for military might.

We have all felt fear
When the guns come near
But we'll not blame you.
Our aim will be true
From the aircraft we hold so dear.

For years we've been told
Of heroes of old
We now feel proud too
To join in your "few"
Of brothers who dared to be bold.

It's comfort to know
You support us so
On missions each night
It helps us make light
Of worries we feel such don't show.

We feel a strong bond
And feel very fond
Of air crew past by
Who've all lov'd to fly
Through the clouds, blue skies and beyond.

To those who have fell
We still toll the bell
And those that now fall
We will miss them all
And meet on heaven not hell.

Jasper Miles.

RIVER MEUSE BRIDGES, SEDAN AREA, 14 MAY 1940

'You British are mad. We capture the bridge early Friday morning. You give us all Friday and Saturday to get our flak guns up in circles all round the bridge, and then on Sunday, when all is ready, you come along with three aircraft and try and blow the thing up.'
German captor to Plt Off I.A. McIntosh on his capture after a vainglorious bombing raid against the bridges at Maastricht.

'We had been standing by at Amifontaine airfield near Reims from 06:00, when at about 15:00 hrs on 14 May 1940 a warm and sunny day, we were called to the Operations hut for briefing on the new threat developing against the Allied front in France. After four days the main thrust of the German offensive had moved from Holland and Belgium to further south towards Sedan on the River Meuse. The operation to be mounted was to meet an appeal by the French General Billotte to AM A. Barratt, C-in-C RAF in France, for a maximum counter-effort by the Advanced Air Striking Force (AASF). The operation was designed to stop the German breakthrough by destroying their pontoon bridges across the river. The briefing emphasised the serious military situation and the importance of the operation. The attack was to be made by approaching at 6,000 ft and dive bombing the target, the four 250 lb bombs fitted with instantaneous fuses being released at 2,000 ft, then returning if conditions allowed at low level.

'12 Squadron's contribution to this operation was five Fairey Battle aircraft, three from "A" Flight and two from "B" Flight, the maximum effort available after recent heavy losses. Most of us had in mind a similar briefing only two days before, on 12 May, for the operation mounted against the bridges at Maastricht. 12 Squadron had been selected to carry it out, resulting in the loss of four of the five aircraft despatched, the fifth being written off at base through battle damage; and the first Air VCs of World War Two being awarded to Fg Off Donald Garland and Sgt Tom Gray, pilot and observer respectively leading "B" Flight in P2204 PH-P. [LAC L.R. Reynolds, the WOP/AG, received no award].

'After the briefing I went to my aircraft – the Wireless Operator then being responsible for the radio and electrical installations – and re-checked the equipment (including the Vickers K gun at the rear) for which I had earlier signed the aircraft's Form 700. My pilot was Sgt H.R.G. "Reg" Winkler, an ex-apprentice coppersmith who had entered the RAF in 1930. The Observer was Sgt Maurice "Bish" Smalley, a direct entry scheme entrant in 1938. L5188 had only recently been returned to the squadron after repair and its new port wing was still uncamouflaged. It was on the night of 1 April 1940, whilst at practice camp at la Salanque in southern France, that with Reg Winkler piloting, we hit an unlit Blenheim while taxiing for take-off. As a result L5188 required a new port wing and propeller, the Blenheim receiving similar damage.

'The "A" Flight target was a pontoon bridge at Donchery west of Sedan and, for "B" Flight, a bridge to the south of the city. At 15:30 hrs "B" Flight took off followed closely by "A" Flight, each flying its own divergent course to the target some 50 miles away. Before reaching Sedan "B" Flight was intercepted by Bf 109s and both aircraft were quickly shot down – one crew member surviving from each aircraft. Fg Off G.D. Clancy, the pilot of L4952, was made PoW. Observer Sgt Alderson and WOP/AG LAC Ainsworth were killed. Sgt A.C. Johnson, pilot of P2322, was killed. His Observer, Sgt E.F.

"12 Squadron's contribution to this operation was five Fairey Battle aircraft . . . the maximum effort available after recent heavy losses." (Author's collection)

White, jumped from his blazing aircraft without a parachute – he was unable to reach it in its stowage among the flames. After being captured, Wireless Operator Fred Spencer had the grim task of identifying the body.

'In "A" Flight (L4950 flown by Fg Off E.R.D. Vaughan, leading, L5538, piloted by Plt Off J.J. McElligott, at No. 2 and L5188 at No. 3) we climbed steadily to 6,000 ft, the signs of war beneath us much of it due to Luftwaffe bombing. Bombed trains, burning buildings, halted road convoys meeting streams of refugees showed the effectiveness of the Luftwaffe's operations. (Our own airfield, a cornfield the previous summer, our needs being met by tents, dugouts and a Nissen hut – had luckily escaped direct attack, the Luftwaffe's attention being concentrated on a small French airfield, La Malmaison, just across the road equipped with a few elderly light biplanes. Usually during the regular visits paid by the Luftwaffe we could see the bombs being released as the aircraft flew directly above, the crump and smoke of bursting bombs being heard and seen for most of the day as delayed-action fuses operated).

'When approaching Sedan we ran into concentrated flak just as Bf 109s were coming up behind. L5538 was hit by flak and was last seen, after jettisoning the bomb load, going down trailing smoke. However, it managed to reach base, the only aircraft to return to Amifontaine, but the crew were killed five days later on 19 May. In a short time L5188 began suffering damage, several holes appearing on the wing upper surfaces although luckily the fuel tanks seemed to escape. One shell passed through the starboard wing between the bombs and flare racks, exploding above. On reaching the target we followed the leader into a dive, Reg pushing the nose down very steeply, so much so that I queried over the intercom whether he was all right. Reg gave me an affirmative. On pulling out I could see dust and smoke straddling the river. Houses were already burning at one side of the river by the old bridge.

'At about 2,000ft when we were again in concentrated flak the leader received a direct hit, pieces of aircraft falling away as it dived to the

"At about 2,000 ft, when we were again in concentrated flak, the leader received a direct hit, pieces of aircraft falling away as it dived to the ground. Only the wireless operator survived." (Author's collection)

ground. Only the Wireless Operator survived. Soon after this we received a direct hit in the engine, a heavy thud which threw the aircraft upwards into Reg's cockpit. With a failed engine and on fire the aircraft, although still controllable, was rapidly losing height. I asked Reg whether he could get it down but at once he gave the order to bale out. By this time he was being forced into a bottomless pit before being jerked into silent downward flight. Looking across I could see the smoke trail left by our aircraft and two parachutes close to the ground in the distance. Shortly afterwards I almost dropped into the canal leading from the River Meuse, all my weight going on my right leg as I hit the canal towpath, my left leg sinking deeply into soft ground. As I gathered myself together, a dozen Bf 109s flew overhead. It was about 16:00 hrs.

'A German motorcycle combination was fast approaching and, feeling rather ludicrous, I put up my hands as directed by the two Germans holding Tommy guns. They searched me, removing my belongings and expressed surprise that I had no pistol. To me the whole situation seemed unreal. I felt detached from the scene as an onlooker watching a drama played out – I couldn't really believe that I was involved. I was put in a sidecar and we drove for about half a mile to a cafe filled with German troops. All along the road tanks were sheltering under the roadside trees.

'I was the first British airman the Germans had encountered and naturally they regarded me with some curiosity, their attitude in the main being friendly and surprisingly sympathetic about my predicament. I had a cut on the back of my head which they bandaged and I was plied with cigarettes, fruit and some coffee. The English-speakers introduced me to that familiar phrase of later days: "For you the war is over." Many laughed, exclaiming that I would not be a prisoner for long because the war would soon be over – how nearly right they were.

'During this time a Hurricane flew over and the Germans let rip with their rifles. The pilot, recklessly I thought, did a few aerobatics at low level, for which the Germans cheered, then flew off. Soon after a Fairey Battle of 218 Squadron, who lost 10 of the 11 aircraft they despatched that afternoon, came streaking past at ground level on full boost, smoke pouring from its exhausts. The aircraft showed signs of damage and the rear position was unmanned. The tanks and other guns opened up and surrounded the aircraft with shell bursts but it disappeared over a ridge. A few seconds later a column of smoke shot into the air and the sound of an explosion followed a few seconds later. The Germans shrugged and exclaimed "est is krieg".

'Eventually my personal belongings, watch, cigarettes, lighter, flying helmet, goggles and parachute ring, were returned to me and a 21-year-old Unteroffizier introduced himself, said he was my escort, and that he had orders to shoot should I attempt to escape. My right leg was by this time very painful. We started walking along the road, the troops having said goodbye, wishing me luck. We walked for probably 30 minutes, the exercise easing the pain in the leg, until we came to a small village. At the end of the street we saw a dying French soldier, a cigarette in his mouth. He had been manning an anti-tank gun which received a direct hit. Another Frenchman was holding him in a sitting position while a German medical orderly attended to his terrible injuries. German infantry was moving through the main street and cattle and a horse in full harness were running aimlessly up and down the road. We were about halfway through the village when shells started falling, the explosions being followed by the characteristic whistle as sound caught up with the trajectory. The troops dropped into the wayside ditches and my escort suggested that we sheltered in the front entrance of one of the houses. The house opposite us received a direct hit and a shell splinter spun across the road and hit my flying boot. I picked it up, it was very hot, and put it in the pocket of my Sidcot. My escort decided it would be safer to join the troops in the ditches. This we did, and then followed an incident which often crosses my mind. I got down and lay alongside a German soldier who turned to me and said in perfect English: "Haven't you got a helmet?" I said: 'No I haven't." He replied: "Get underneath me", and moved over. I did so and lay with this soldier sheltering me. The shelling ceased about five minutes later and the troops started getting out of the ditches. I got up and thanked the German. He said it was nothing, that I was lucky, I would survive the war, but for him and his comrades, no-one knew what the future held. He wished me luck and a safe return to my home. And so we parted. I have often wondered about this German. Who was he? What was his motivation? Did he survive?

'After some time we came to a local HQ and my escort, after wishing me well, handed me over to some officers who invited me to enter a staff car. By this time it was dusk. We drove for about half an hour, our way impeded by a continuous jam of vehicles attempting to move in the opposite direction. We arrived at a large house not far from the River Meuse, which appeared to be some kind of HQ. On leaving the staff car I left behind my helmet, goggles, and parachute ring, not realising I had done so until next day.

'I was taken to an armoured vehicle in which sat a senior officer studying maps at a table. A wireless operator was working in a corner. The officer, who spoke perfect English, asked me about the wound on my head, congratulated me on surviving my action and asked where I had come from, to which I replied: "England". He asked if I had any complaints about the treatment given me and said that if any of my personal property had been taken he would take steps to get it back. He mentioned that in peacetime he was a regular visitor to England and had enjoyed the Shakespeare season at Stratford the previous summer. He said goodbye and suggested that I should have a meal in the house. About a dozen sat at the table lit by a couple of oil lamps, but I was not hungry. I was drinking some beer when a heavy explosion occurred nearby, rocking the house as an aircraft passed very low overhead. We all dived for the floor and dust and some debris fell around. On picking ourselves up my escorts suggested we should move on.

'We walked for a little way and came to the river where a number of rubber boats were being used to ferry troops across. We were standing by the bank when tracer was seen a few hundred yards up the river. A mobile searchlight switched on, illuminating a Blenheim. A two-man-operated machine-gun opened up and tracer criss-crossed the aircraft but it flew on. The aircraft's gunner probably changed a pan at this time, because firing ceased as he passed us at about 200 ft and then continued when further along the river. We got into one of the boats, which was paddled to the other bank. We then walked for some time along a road jammed with stationary vehicles until we came to a farm. I was handed over and taken to a farm building full of French soldiers. By this time it was midnight. I sat on the straw, my head swimming with exhaustion, sundry aches, cuts and bruises forgotten, and went to sleep. It had been a long day.

'And so ended the day, an operation resulted in the RAF suffering its heaviest loss ever in the war for an operation of comparative size – 40 of the 71 bombers despatched being shot down. And what was achieved? Sadly, very little. Two pontoon bridges destroyed and two damaged with very little effect on the German advance. Of the 40 crews lost, 17 members made their way back from enemy territory. The other 103 were either dead or PoW.

'Next morning I met up with Reg (his "singed" neck bandaged) and Bish Smalley. We journeyed through the Ardennes in a lorry with a stay overnight in a church, a few days in a warehouse at Libremont in Belgium and rougher treatment as we got further away from the front. A two-day coal truck journey to Stalag IVB Muhlsburg, then eventually to Dulag Luft, Limburg, Lamsdorf, Barth, Sagan, Hyedkrug, Thorn, and Fallingbostel.

AC1 Len R. Clarke, WOP/AG No. 12 (Bomber) Squadron, 1940.

The Ballad of Sulaiman

In the year anno domini one-ninety-four
Was just outside Sulaiman there started a war
HQ got excited and set down to work
To pull operations stuck out in the dirt
Chorus: *No bombs at all, no bombs at all,*
If our engines cut out we'll have no bombs at all.

There were once two pilots sent out to bomb Sul,
Their bombs were all right but their tanks were half full
Then from the back came the agonized call,
"If our engines cut out we'll have no bombs at all".

T'was just over Sul and both engines cut out;
And again from the back came the agonized shout:
"If we land to the east of the Basrian Pass
Might as well stick that Lewis Gun right up our arse."

They looked o'er the side and t'was a quite plain to see,
That all the hill tribesmen were seated at tea.
And sitting around 'midst their herds and the rocks,
Discussing spring fashions in stockings and socks

St Peter reclined on a large fleecy cloud
When the Orderly angel came fluttering round.
"Excuse me, St Peter, Allow me to say
That an old Rolls-Royce Merlin is coming this way.

They landed at last and were full of good cheer
St Peter said: "Chaps, shall we split the odd beer?
The pilots replied in a voice loud and shrill:
"We thank you, St Peter, we think that we will."

The moral of this it is quite plain to see
Look after your fuel tanks where'er you may be,
And when 'midst the enemy you must roam,
If you must employ petrol, don't leave it at home.

THE SONG OF THE MERLIN

Touch me gently, wake me softly,
Let me start to sing,
Free to use my strength and power
Throbbing on the wing.

Let me roar my throaty war-song
As we start to rise,
Challenging the unseen dangers
Lurking in the skies.

Keep me happy as we settle
Steadily to fly,
Purring like a drowsy kitten
Loudly in the sky.

Let me never fail my masters
In the searchlight's glare,
Let me keep them safely airborne
In my loyal care.

Let me help them do their duty
On their awesome flight,
Let me bring them through the hazards
of the savage night.

Winging through the cloudy darkness
I will sing my song,
Reaching up to dawn's pale sunlight
Though the night was long.

Let me sing of men returning
Safely homeward bound,
Then my thrusting heart shall sing
In glorious joyful sound.

Audrey Grealy

Nationalities of Pilots and Aircrew who Flew Operationally Under Fighter Command Control in the Battle of Britain, 10 July–31 October 1940
(Nos killed in parenthesis)

RAF (British)	2,365	(397)
Fleet Air Arm	56	(9)
Australia	21	(14)
New Zealand	103	(14)
Canadian	90	(20)
South Africa	21	(9)
Southern Rhodesia	2	(0)
Jamaica	1	(0)
Ireland	9	(0)
USA	7	(1)
Poland	141	(29)
Czechoslovakia	86	(8)
Belgium	29	(6)
Free French	13	(0)
Palestine	1	(0)
Total	**2,945**	**(507)**
Wounded	500	

Losses, July–November 1940

RAF pilots	975	killed
	443	wounded
RAF aircraft	925	destroyed
	343	damaged
Luftwaffe aircraft	1,537	destroyed

'The English have lost the war, but they haven't yet noticed it; one must give them time, but they will soon come around to accepting it.
Adolph Hitler, 17 June 1940, on the Fall of France.

'What General Weygand called the Battle of France is over. I expect that the Battle of Britain is about to begin. Upon this battle depends our own British life, and the long continuity of our institutions and our Empire. The whole fury and might of the enemy must very soon be turned on us. Hitler knows that he will have to break us in this island or lose the war. If we can stand up to him, all Europe may be free and the life of the world may move forward into broad, sunlit uplands. But if we fail, then the whole world, including the United States, including all that we have known and cared for, will sink into the abyss of a new dark age made more sinister, and perhaps more protracted, by the lights of perverted science. Let us therefore brace ourselves to our duties, and so bear ourselves that, if the British Empire and its

Top-scoring Pilots, RAF Fighter Command, July-November 1940

Name	*Nationality*	*Squadron*	*Aircraft*	*Total*	*Remarks*
Lock, Plt Off E.S.	British	41	Spitfire	22 + 1 shared	WIA 17.11.40
Lacey, Sgt J.H. 'Ginger'	British	501	Hurricane	18	
McKellar, Flt Lt A.A.	British	605	Hurricane	18 + 1 shared	KIA 1.11.40
Frantisek, Sgt J.	Czech	303	Hurricane	17	K 8.10.40
Carbury, Plt Off B.J.G.	New Zealand	603	Spitfire	15 + 1 shared	
Doe, Plt Off R.F.T.	British	234/238	Spitfire/Hurr	15	
Urbanowicz, Sqn Ldr W.	Polish	303	Hurricane	15	
Hughes, Flt Lt P.C.	Australian	234	Spitfire	14 + 3 shared	KIA 7.9.40
Gray, Plt Off C.F.	New Zealand	54	Spitfire	14 + 2 shared	
Crossley, Sqn Ldr N.M.	British	32	Hurricane	13 + 2 shared	

Commonwealth last for a thousand years, men will still say: "This was their finest hour".'
Winston Churchill, 18 June 1940.

'It would be the bankruptcy of statesmanship to admit that [the bomber] is a legitimate form of warfare to destroy its rival's capital from the air.'
The Times, 1933.

'I was visiting a big aircraft factory the other day and I was talking to a crowd of workers. I was not quite sure how they were reacting to my remarks, so I turned to a little old man who looked particularly surly and said to him: "How long do you think you can stick this war?" Without hesitation the chap said: "One week longer than the bloody Germans".'
Lord Trenchard, June 1940.

'I have told the Fuhrer that the RAF will be destroyed in time for Operation Sea Lion to be launched on September 15, when our German soldiers will land on British soil.'
Hermann Goering, 6 August 1940.

BIG WING

At Duxford on 1 July 1940, 19 Squadron's strength, officially, was only eight aircraft (with five unserviceable). On the same date 264 Squadron had 11 Defiants available. It was the same story everywhere. To meet the massive threat posed by the all-conquering Luftwaffe in what would become the Battle of Britain, RAF Fighter Command numbered just 57 squadrons. They were equipped with Hurricane Is, Blenheim IFs, Spitfire Is, and Defiant Is spread throughout three Groups: 11 Group, which would be in the

At Duxford on 1 July 1940 19 Squadron's strength, officially, was only eight aircraft (with five unserviceable). (via Andy Height)

front line, with 29 squadrons; 12 Group, (of which Duxford was one of its five airfields), with 11 fighter squadrons equipped with only 113 serviceable Spitfires, Defiants, Hurricanes and Blenheim IFs; and 13 Group, which covered the north-east, with 17 squadrons.

On 10 July 310 (Czech) Squadron was formed at Duxford under the command of Sqn Ldr G.D.M. 'Douglas' Blackwood with Hawker Hurricane Mk I aircraft. The Czechs became operational on 17 August with a patrol along the Thames Estuary. On 26 August they tussled with Bf 110s, losing three Hurricanes for the destruction of one Bf 110. All three Hurricane pilots were saved. (A sister Czech Hurricane Squadron, 312, was formed at Duxford on 29 August, but did not become operational until 2 October).

The Battle of Britain was reaching its height.

At Duxford the Station Commander, Gp Capt A.B. 'Woody' Woodhall, the fighter controller, ordered: '242 Squadron scramble! Angels 15. North Weald.' (Author)

On Friday 30 August the Luftwaffe began a 48-hour assault on Fighter Command's Sector stations. Number 11 Group was threatened with being overrun by sheer weight of numbers, and 12 Group now had to act as an airborne reserve for their hard-pressed colleagues. Early that morning 242 Squadron, commanded by Sqn Ldr Douglas Bader at Coltishall, Norfolk, was scrambled but was recalled while en route to Duxford. Ordered off again, the Hurricanes were in position by 10:00 hrs. At Duxford the Station Commander, Gp Capt A.B. 'Woody' Woodhall, the fighter controller, ordered '242 Squadron scramble! Angels 15. North Weald.' Bader ignored the request to 'Vector one-nine-zero. Buster', flying 30° further west to get up-sun. When he spotted 50 Dorniers escorted by Bf 110s bearing down on North Weald, Bader led 242 down into the German formation and the Hurricanes routed them. Twelve enemy bombers were shot down and North Weald escaped destruction but the Vauxhall Motor Works at Luton was badly hit and 53 civilians killed and 60 injured.

Altogether, the day's fighting cost the RAF 26 fighters shot down, and the Luftwaffe 36. Back

A series of prints by Ralph Hull depicting the Battle of Britain

at Coltishall Bader was congratulated by the AOC 12 Group, AVM Sir Trafford Leigh-Mallory. Bader argued that with more fighters they could have shot down three times the number. In theory, a large fighter formation could be brought to bear on the enemy 'Balbo', thereby increasing the chances of 'knocking down' more aircraft than smaller formations were capable of doing. Leigh-Mallory had long held the belief that a 'Wing' could achieve greater killing potential than the squadron formations favoured by ACM Sir Hugh 'Stuffy' Dowding at Fighter Command and by AVM Keith Park at 11 Group, who had much less time to form up squadrons into Wings.

On 31 August 242 Squadron were scrambled three times to patrol North London once more, but they found nothing. Fighter Command lost 39 fighters and Debden, Biggin Hill, Manston, West Malling, Hawkinge, Hornchurch, and Lympne got a pasting from Luftwaffe bombers. Duxford was only saved by Sqn Ldr J.M. 'Tommy' Thompson's 111 Squadron from Debden. South of the Thames only two RAF Sector stations were still operational. Leigh-Mallory landed at RAF Coltishall and talked

Next morning 242 Squadron flew to Duxford, where Bader (centre) and his pilots spent a frustrating day waiting in vain to be summoned by 11 Group. (via Frank Crosby)

'Wings' to Bader between patrols. The commander told Bader that, starting on the morrow, 242 and 310 Squadrons' Hurricanes would use Duxford daily. Together with 19 Squadron operating out of the satellite at nearby Fowlmere, they would form the 'Big Wing'.

At first there was no 'trade' for Bader to pursue. He practised with the 'Duxford Wing' for four days and reduced take-off times to just three minutes, the same as a squadron, but they were not called into action. On 5 September 19 Squadron, still stationed at Fowlmere, lost three Spitfires and the CO, Sqn Ldr P.C. Pinkham, was killed. Sqn Ldr B.J. 'Sandy' Lane DFC took over command. On 6 September six of 11 Group's seven sector stations and five of its advanced airfields were very badly damaged. With the sector stations expecting annihilation, and a signal being sent that an invasion was 'imminent', on 7 September the Luftwaffe switched its daylight attacks, sending 300 bombers to hit London. Woodhall asked Bader to 'Orbit North Weald. Angels ten'. He went instead to 'Angels 15', but was still below the enemy formation when it was sighted heading for the capital. Although 19 and 310 Squadrons never did catch up with 242 Squadron, 'Bader's Bus Company' claimed 11 enemy aircraft for the loss of two Hurricanes, though one pilot was safe.

Next morning 242 Squadron flew to Duxford, where Bader and his pilots spent a frustrating day waiting in vain to be summoned by 11 Group. The same thing happened on the 9th, until at 17:00 hrs it was announced that radar had detected a build-up of German aircraft over the Pas de Calais. Only when the bombers headed in were the Hurricanes permitted to scramble. Woodhall asked Bader, 'Will you patrol between North Weald and Hornchurch, Angels 20?' Bader disregarded this and climbed south-west to 22,000 ft. When he sighted the bombers he ordered 19 Squadron's Spitfires to climb higher and provide cover as the Hurricanes attacked in line astern through the middle of the enemy bomber formation. The 'Duxford Wing' routed the bombers and claimed 11 of the 28 officially destroyed this day. Three Hurricanes of 310 Squadron failed to return. At 17:30 hrs Fg Off Gordon Sinclair baled out and landed in Caterham High Street after colliding with Plt Off J.E. Boulton, who also hit a Dornier and was killed. The third man lost, Plt Off Rypl, was unhurt after crashing near Oxted out of fuel. At 17:45 hrs Plt Off K.M. Sclanders' and Sgt

On 13 September Bader . . . was given two more squadrons (302 Polish with Hurricanes and 611 with Spitfires) to bring the wing up to 60 fighters. (BAe)

R.V.H. Lonsdale's Hurricanes in 242 Squadron were shot down. Lonsdale baled out, but Sclanders was dead.

On 13 September Bader, who was awarded the DSO, was given two more squadrons (302 Polish with Hurricanes, and 611 with Spitfires) to bring the Wing up to 60 fighters. Next day the two squadrons joined 242, 310 and 19 Squadrons at Duxford and Fowlmere. Twice the 'Duxford Wing' patrolled North London; to no avail. On Sunday 15 September the first wave of Luftwaffe aircraft plotted heading for London were engaged by 11 Group. When the next wave came in the five-squadron wing was summoned to action by 11 Group, now badly in need of reinforcements. Woodhall asked Bader to patrol Canterbury-Gravesend. He did, but though Sandy Lane's and 611 Squadron's Spitfires assailed the bombers, Bader's and the two other Hurricane squadrons were jumped by Bf 109s.

The Wing landed, refuelled, and was at readiness again by 11:45 hrs. Two hours later the Wing was scrambled again. At 14:15 hrs Flt Lt G.S.F. Powell-Sheddon was shot down by a Do 17 but was safe. Bader still complained of being scrambled too late. Arguments that 'Big Wings' were unwieldy seemed justified, especially when 11 Group's own three-squadron 'Wing' of 32 fighters, led by Bob Stanford-Tuck from Debden, had only eight fighters remaining by the time they intercepted the bombers. The 'Duxford Wing' claimed 52 enemy aircraft shot down in the two engagements and eight probably destroyed. Overall, Fighter Command claimed to have shot down 185 aircraft, but the true figure was 56 German aircraft shot down for the loss of 26 RAF fighters. Three came from 19 Squadron, 310 Squadron lost two Hurricanes, and 611 one Spitfire.

On 18 September the Wing claimed 30 destroyed, six probables and two damaged; 242 Squadron claimed 11 of these for no loss and 19 Squadron lost two Spitfires, though their pilots were unhurt. The last 'Big Wing' 'thrash' came on 27 September, when the Luftwaffe lost 55 aircraft, the RAF losing 28 fighters. Both 242 Squadron (Plt Off M.G. Homer was killed) and 302 Squadron lost one Hurricane, and 19 Squadron lost two Spitfires. Gordon Sinclair in 310 Squadron was shot down again but was unhurt. Total claims by the 'Duxford Wing' for 27 September were 12 destroyed, bringing Bader's Big Wing claims to 152 aircraft shot down for the loss of 30 pilots. 'Big Wing' controversy still reigned. In the aftermath of the

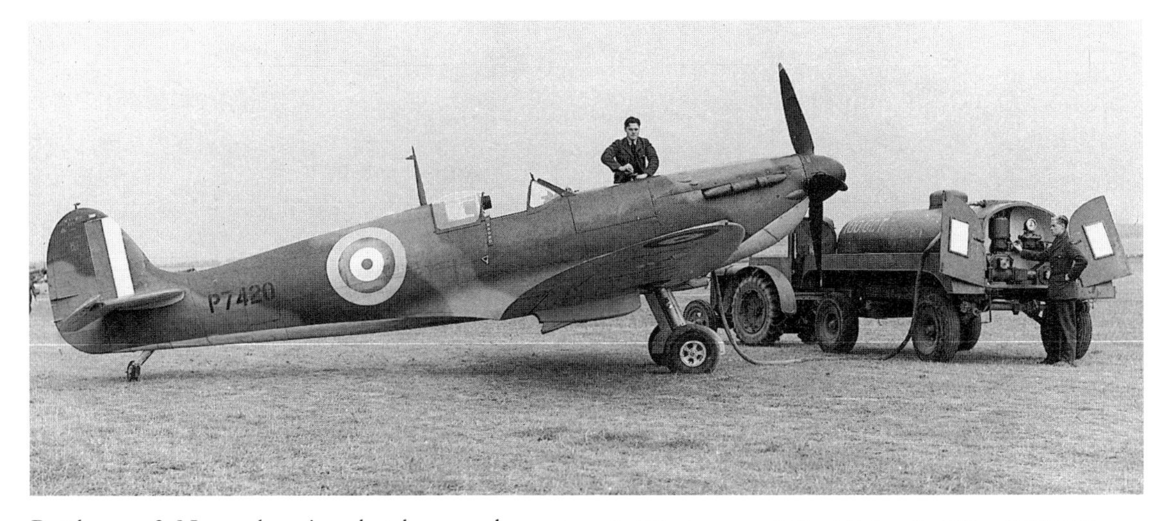

Sgt Roden was killed flying Spitfire II P7420, seen here at Duxford, when he flew into a tree at Boxford, Sussex, on 15 November 1940. (via Frank Crosby)

Battle, on 2 November, in what became known as the 'Duxford memorandum', Capt Harold Balfour, Undersecretary of State for Air, recorded the arguments for and against the 'Big Wing' put forward by 11 and 12 Groups respectively. As the RAF turned to the offensive, the architects of victory, ACM Sir Hugh Dowding at Fighter Command and ACM Sir Keith Park, AOC 11 Group, handed over to ACM Sir W. Sholto Douglas and ACM Sir Trafford Leigh-Mallory respectively. In March 1941 Sholto Douglas established the post of wing commander flying at all main airfields, and Bader was given the Tangmere Wing.

'The gratitude of every home in our island, in our Empire and, indeed, throughout the world, except in the abodes of the guilty, goes out to the British airmen, who, undaunted by odds, unweary in their constant challenge and mortal danger, are turning the tide of the world war by their prowess and their devotion. Never in the field of human conflict was so much owed by so many to so few.'
Winston Churchill, 20 August 1940.

ERKS

I didn't want to join the Air Force,
I didn't want my bollocks shot away.
I'd rather hang around
Piccadilly Underground,
Living off the earnings of a high-born lady.

'Don't trust the erks to clean your Perspex surfaces: check them yourself, during daylight, before an operation. Plt Off Prune, a cartoon character in the RAF's surprisingly light-of-

(Tee Emm)

ORDER BY POST WITH CONFIDENCE WE HAVE A LARGE & EFFICIENT STAFF TO GIVE EVERY ATTENTION TO YOUR REQUIREMENTS

FREE INSURANCE
given with every copy covering against Accident or Mishap—if you've read and understood the gen.

(Tee Emm)

Pilot Officer Prune says —
"Take* Tee Emm *regularly!
Prevents that Thinking Feeling!"

touch training magazine, *Tee Emm*, was once shown hanging by his straps during violent evasive action occasioned by a spot on his cockpit canopy which he had mistaken for a Hun in the Sun.'
Graham Harrison, gunner.

'In June 1940 I went to the RAF Recruiting Office in Kingsway, London. I was 18 years old and had been a photographer in civvy street. The pilot officer asked me what I wanted to do. I said I wanted to fly aeroplanes. (I was wearing glasses because I was short-sighted, and had been since the age of six). He told me that because of my poor eyesight I would have to consider my responsibility to other crew members. He said: "We can't let you fly."

'I said: "What responsibility to other crew? I want to fly Spitfires or Hurricanes!"

'He said that if a bullet shot my glasses off, they would lose a valuable aircraft. So I said that if I had perfect vision and a bullet came that close, they would lose a valuable aircraft any way.

'So that got me nowhere. The pilot officer said: "Well, what would you like to do in the RAF on the ground?"

'I said I would like to be an aerial photographer. To which he asked: "What do you understand by a circle of confusion?"

'I said: "As far as I understood it was three pilot officers discussing a technical point!"

'They made me an electrician.
Dennis 'Joe' Davis, 515 Squadron.

'ODDS AND SODS'

'AC2 plonk was the lowest form of animal life in the Service, they used to say. I was 20 years old, doing guard and any odd jobs. To my

(Gordon Kinsey)

(Gordon Kinsey)

amazement one afternoon in 1940 these damn great lumbering Whitleys suddenly started arriving en route to bomb the Skoda Works at Pilsen. We understood that they had come from Yorkshire. The old sweats called them "flying coffins". The crews went off to the mess. In the evening, a flight sergeant with an Aldis lamp gave them the go-ahead to take off. The first time I saw them start up I thought: "The damn things are on fire!" I forgot then that you primed the Merlin engines and damn great flames kept shooting out. They got going, took off one after the other, and out into the night.

'When they returned, to our amazement they put their landing lights on and they illuminated the church. Considering that we were used to the blackout the light was enormous. Eventually, the crews drifted out to the pub in Dereham. Now the road was being repaired. These clever blokes, I suppose they wanted a lift back, took all the "Diversion" and "Road Blocked" signs and put them right across the Norwich-Dereham road. The road was diverted past the camp. And of course they all got nice lifts back! I was on guard and heard some of the blokes singing a song:

'"Goodbye to to Swanton, goodbye to you,
We've strolled around your town, screwed your women too.
Your NAAFI beer is lousy, operations are a farce,
And you can stick your service flight, stick it up your arse!"

'I said, "Halt, who goes there?", and they replied "**** off!"

I was completely nonplussed.'

Pat Holmes, RAF Swanton Morley

SPROGS

We had been flying all day long at a hundred effing feet,
The weather effing awful, effing rain and effing sleet.
The compass it was swinging effing south and effing north,
But we made an effing landfall in the Firth of effing Forth.

Chorus: *Ain't the Air Force effing awful? (repeat twice)*
We made an effing landfall in the Firth of effing Forth.

We joined the effing Air Force 'cos we thought it effing right,
But don't care if we effing fly or if we effing fight.
But what we do object to are those effing Ops Room twats,
Who sit there sewing stripes on at a rate of effing knots.

'Sprog crews, for their few operations, were assigned to any spare aircraft available and we were to do six ops before being allocated our "own" aircraft, namely a Halifax BIII with Hercules XVI engines, number LV907 and lettered "F". So many aircraft with the letter "F" had been lost that it was felt that an unlucky name for this one would not make any difference anyway, and it was "Jumbo" Smith, the pilot, who handed over to us on completion of his tour, who named it *Friday 13th*. This name, plus skull and crossbones, an upside-down horseshoe and the words: "As Ye Sow, So Shall Ye Reap," were painted on the nose of the fuselage, with a

"Dummy pops off and attacks the docks at Brest — OK?"

(Tee Emm)

(Tee Emm)

"And with the throttles closed and stick held back, the aircraft will sink gently on to three points."

"bomb" to represent each completed operation.

'I think most of us harbour some superstition, although not always ready to admit it, but I had miraculously survived Course No. 13 at at Mesa, Arizona, home of No. 4 British Flying Training School, better known as Falcon Field. *Friday 13th* had safely completed a total of 42 operations by the time we took it over, and so our delight in having our own aircraft must have outweighed any doubts we may have felt.

'Again the magic No. 13 had proved lucky for me as we flew 26 operations on *Friday 13th* to complete our tour. We handed over to "Doc" Gordon, a RCAF pilot and his crew, who were to

In the early part of the war Blenheim crews (Flt Sgt Fred Thorogood, WOP/AG, far left) did almost suicidal low-level sweeps over the Continent and against enemy shipping. (Ralf Thorogood)

complete their tour on it, and by the end of the war in Europe a few other crews took it to the quite remarkable total of 128 full operations. It was then exhibited in Oxford Street, London, before being scrapped. An Australian lad attempted to purchase the aircraft, but he was too late. He did, however, manage to obtain the fuselage panels bearing the name and bombing tally, and these remained under his house until 1972, when they were flown back to the UK for display in the RAF Museum, Hendon, where they still remain in a glass case, together with our crew photo.

'So much for superstition?'

Derek Waterman, DFC, pilot, 158 Squadron.

'In July I received instructions to report to RAF Padgate, near Warrington, part of which was a reception for new recruits. Padgate was a most forlorn and forbidding looking place. I reported to the Guard Room and was given a hut number and directions and told to report to the corporal in the charge, who allocated me a bed in a very large hut with about 50 beds in it. The bed was very austere and consisted of three "biscuits" which when put together formed a mattress, a pile of blankets, and a rather dirty looking pillow – no sheets! There were about 30 men in the hut at the time, all considerably older than myself, who were just sitting about on their beds smoking and chatting to one another. Eventually I went to the ablutions. I left my sponge bag on a wash basin and that was the last I saw of it! What an introduction to the RAF! After visiting the NAAFI I returned to the hut and prepared for bed. When I put on my pyjamas there was a howl of derision from the other members of the hut and for a moment I was unable to establish the cause, but then realised that pants and vests were *de rigueur* and pyjamas were regarded as something of an oddity, if not worse.

'We were woken up at the crack of dawn by our corporal who seemed to be in a foul temper, and after a perfunctory wash we were marched off to the Airmen's Mess where we queued up for breakfast, an austere meal.

'At about 9 o'clock we were mustered outside the hut and marched to the equipment stores to be issued with all our kit. It was a pretty formidable list consisting of uniform, battledress, shirts and collars, tie, socks, vests, pants, boots, a button stick, knife, fork, and spoon – referred to as "irons", gas mask, a "housewife" (pronounced "huzif"), which was a sewing kit, and finally a huge white canvas kit bag, the packing of which was an absolute art which took some time to acquire.

'When we returned to the hut we were told to put on our battledress and pack up all our civilian clothes, which would be sent to our home address. The battledress trousers were so coarse that I wore pyjama trousers underneath.

'We were posted to Blackpool for basic training. We marched to the railway station carrying our kitbags, gasmasks and greatcoats, which was a formidable load. On arrival the air crew trainees were separated from the rest of the recruits and taken to a hotel which had been commandeered as the Air Wing HQ. We were quartered in boarding houses, about eight to a house. I was fortunate in having a quite motherly landlady and we were made to feel welcome, which was not the experience of some of the other chaps, whose less hospitable landladies' house rules were a bit severe.

'After breakfast each morning we assembled on the promenade for drill and PT which continued throughout the morning except for a short break for a cup of tea in one of the promenade cafes. All u/t aircrew wore white flashes in their forage caps and were an automatic target for the drill NCOs, who picked on us for the slightest excuse, real or imaginary – ties not properly tied, untidy bootlaces, buttons and boots not polished to perfection, etc. We were paraded morning and afternoon and on one morning the sergeant hissed at me: "'aircut, you". At lunchtime I had it done but at the afternoon parade he stopped at me again and said: "I thought I told you to get an 'aircut". He sent me to have another. Two haircuts in one day was a bit much, but the barber did the second one for nothing!

'When not marching up and down the prom or doing PT, we went to the tram sheds to learn Morse code. Towards the end of the course the footslogging was reduced and the technical training increased. The final examination eliminated a number of would-be operators and those who passed were given a week's leave before reporting to a wartime radio school based at the Albert Hall with quarters in the adjacent Albert Mansions.

'After this we were posted to various operational stations whilst awaiting vacancies at No. 2 Radio School, Yatesbury. I was sent to 103 Squadron at Elsham Wolds in Lincolnshire. I was employed in the W/T section of the squadron and managed to get in a few hours in Wellington

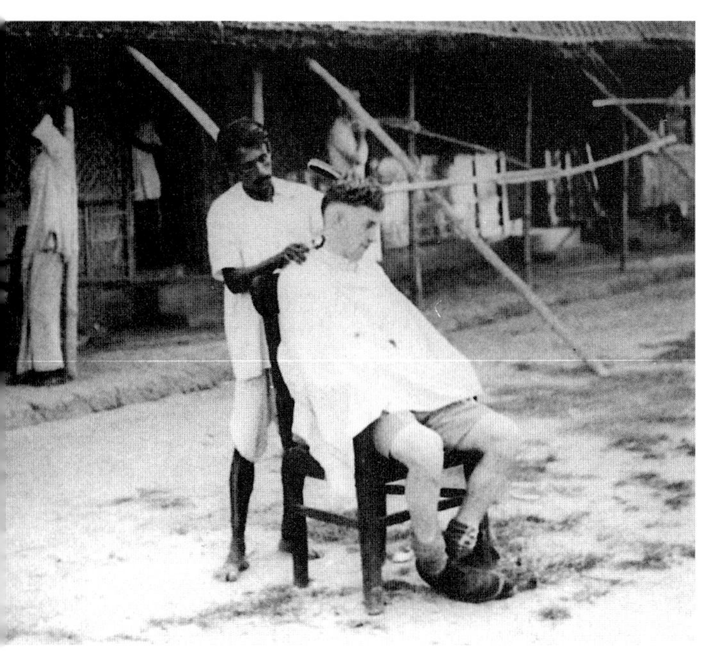

''aircut, you!' (Wg Cdr Jones)

air testing radio equipment. While I was there the first 1,000-bomber raid, against Cologne, took place. To make up the numbers several aircraft and u/t crews were brought in from the OTUs.

'I was at Yatesbury for about six weeks, during which time I completed a variety of wireless exercises, and after passing the final exams I was given the wireless operator's badge. One of my instructors was Sgt Hannah, who had received the VC for crawling out on the wing of a Hampden to extinguish a fire in an engine when returning from a bombing trip to Germany. He was not at all a fit man and later died. After a week's leave I was posted to No. 3 Air Gunner School at Mona, Isle of Anglesey, where we flew on the Blackburn Botha, a hideous aircraft. The course lasted about three weeks, during which time we became familiar with various gun turrets, including the Frazer-Nash. We were also taught how to strip and rebuild the Browning .303 machine-gun, clear blockages, use the reflector sight, gun harmonization, aircraft recognition, and many other bits of useful information. All in all it was a fairly intensive course. We took the final examination, at which there were a considerable number of failures, and at the passing out parade we were addressed by the CO and given our AG brevet and Sergeant stripes.'

Plt Off John Bateman, Boston gunner.

'GET SOME HOURS IN'

'The actual flying conversion to Whitley IIIs was quite a step. A real heavyweight was my first impression, but in practice it was not too difficult to fly once the delayed action control response was accepted. Take-off produced a distinct swerve to port and was something to get accustomed to. A large part of the training was devoted to navigation involving quite long cross-country flights. Cross-country exercises at night at first were quite taxing – it was a different world entirely, finding one's way around in the dark, but at the age of 20 one soon learns and the whole experience becomes second nature. Much of our flying was carried out from a satellite airfield – Stanton Harcourt. There was such a thing as the Harcourt medal, a fictitious gong awarded to the biggest clot of the week, and I have to report that my rear gunner managed to deliver a burst from his four Brownings over the village. Fortunately, nobody was hurt, but apparently the village horse-drawn cart was last seen disappearing at great speed, milk flowing freely from holes in the churn.'

Basil S. Craske, RAFVR, Whitley pilot, OTU Abingdon, Oxon, 1940.

AB INITIO

You may cuss the Tiger Moth, while you're blowing off the froth
From your tankard of good honest English beer, man,
But they put me through my paces in the great wide open spaces
When they trained me on the A.T. and the Stearman.
Yes, I trained out in the States, out in Arizona, mates,
In a place that was as hot as Satan's kitchen,
And it did no good to fret when you fairly dropped with sweat,
'Cause they only tell you, "Brother, quit your bitching!"

David Livingstone-Spence (Course 25), No 4 BFTS, Mesa, Arizona, writing in The Falcon, *the school's magazine.*

(Deacon cartoons, The Falcon, *magazine of No. 4 BFTS)*

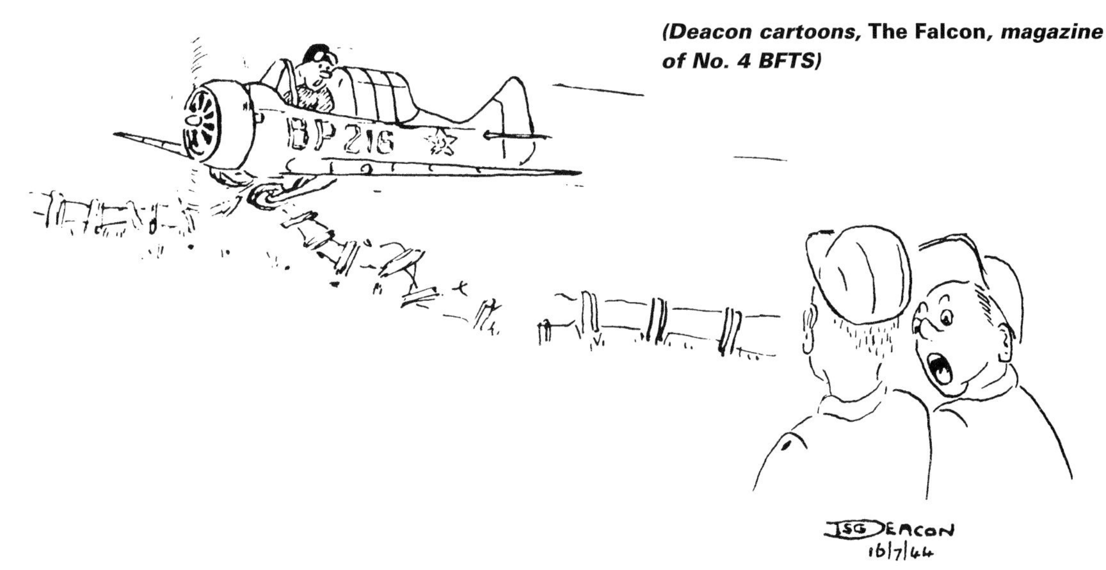

"That's the second time this week he's taken the boundary fence on take-off!!"

You may cuss the Tiger Moth, while you're blowing off the froth. (de Havilland)

'In addition to training schools in the UK, which trained over 88,000 aircrew, a large number of schools were set up overseas under the British Commonwealth Air Training Plan. Canada became the main centre for training, where a total of 137,000 aircrew were trained. The other Commonwealth countries also made a major contribution; Australia trained 27,000, New Zealand trained 5,000, South Africa trained 25,000 and Rhodesia trained 10,000, with other schools being set up in the USA, which produced a further 14,000 aircrew . . .'
Chris Hobson, Senior Librarian, RAF Staff College, Bracknell, writing in 1995.

'Les Jones climbed to 4,000 ft and began his aerobatics in the Stearman. He did not notice a strong wind blowing him north. When he was ready to return to base he was LOST. He looked

(Deacon cartoons,* The Falcon, *magazine of No. 4 BFTS)

No! – I wasn't low flying sir! – it was a* very very *tall cactus!!

for a space to land and finally put down on a small landing strip near Rockwall. Some Texans in stetsons, chaps and wearing revolvers, came over and said: "Hi y'all. Where you from son?" After a brief explanation and directions from the Texans, Les said: "Thanks a lot. I'll get off then."

"Not before you pay the landing fee," said the Texans.

"Landing fee?", said Les. "How much?"

"$25 American, son," said the Texan.

'The outcome was that Les walked some way to a 'phone and called base. Eventually, two instructors flew out in a Stearman, paid the fee, and Les flew back.'

L. James Freeman, cadet pilot, No. 1 BFTS, Terrell, Texas.

'Flying a Stearman properly was more difficult than a Harvard.' (Author)

'After Tiger Moths, which are of course, a lot lighter, I had some difficulty in landing the Stearman. In my opinion, flying a Stearman properly was more difficult than a Harvard. It was more "seat of the pants" stuff. It was a very tough, robust aircraft, and despite a few problems, I did enjoy it!'

Paul Ballance, who completed 70 hrs on Stearmans at No. 1 BFTS, Terrell, Texas.

In all, about 2,200 cadets passed through Terrell. After return to the UK, Nos. 4 and 7 Courses suffering particularly badly on operations. Flight Sergeant Arthur Louis Aaron from No. 6 Course was awarded a posthumous Victoria Cross for his act of courage while a pilot of a Stirling in 218 Squadron, Bomber Command, which made a night attack on Turin, Italy, on 12 August 1943.

'As I was approaching to land I could see another Stearman behind me on the crosswind leg . . . I looked up and back for the other Stearman. All I saw was a propeller somewhere over my tail and two wheels heading my way. I ducked my head low and heard (and felt) a crunch. Those wheels had gone into my upper wing on either side of the fuel tank. At the same instant a fairing strip at the joint of wing and fuel tank zinged like a spring somewhere above my head. At that moment the other 'plane was virtually superimposed on mine. Fortunately, there was a fairly good wind blowing. The other cadet knew exactly what had happened, gave her full throttle and dragged her off. As his tail passed over it also struck my top plane, but he went on to do a normal landing. In a concerned state I hurried over to the dispatch area. Someone said: "If I wuz you I'd get a gun and shoot the sonofabitch!" The cadet was washed out immediately and sent to Trenton, Canada, where they remustered cadets to other aircrew trades.

'Where this cadet's fate was not deserved, another's was. Careless taxying remained a curse in the RAF throughout the war. In this instance a cadet taxying his Stearman back to the flight line headed directly for the wing of the last parked one and proceeded to chew his way through almost half the starboard wings.'

Victor Deboni, No. 1 BFTS, Terrell. More than 7,000 pilots were produced for the RAF at seven British Flying Training Schools in the USA.

'One of my most vivid memories is my first solo in the Stearman. I was having a particularly bad morning doing circuits and bumps with my primary instructor, Ray Shelton, a nice but emotional man who played a violin in his spare time! I had done about 5 hrs on the Stearman and cadets were given an elimination test at 7 hrs if they had not gone solo. So I took the bull by the horns and suggested that I had a go on my own! Ray Shelton stood up in the front cockpit, threw his helmet on the ground and said: "You f****** sonofabitch, you will break your godamned neck."

'I calmly replied: "No I won't." Without further ado, he threw his parachute out of the cockpit and sloped off, muttering "sonofabitch". I got on with the job and managed to get up and down, just avoiding two ground loops. The remainder of my course went very well, including the Advanced on AT-6s (Harvards). I think that my subsequent career in aviation and the "exceptional" assessments in my log books justify Ray Shelton's decision.'

Brian Trubshaw, CBE, MVO, FRAeS, 4 BFTS, Mesa, Arizona, (later Concorde test pilot).

'Perhaps the first thing we noticed, after gazing with awe at the Stearman aircraft on "the line", was the names of some of the flying instructors – Goethe, Schmidt, Burkhalter, Fitze, Haut, and Schellenberger.'

Reg Everson, who arrived at Darr Aero Tech, Albany, Georgia, on 2 October 1941, and who later flew Mosquitos in Europe.

During its existence, 2,000 British cadets passed through Darr. When the Arnold Scheme finished in February 1943, with Class 43-D, the 13 classes had produced some 4,500 RAF pilots from an intake of 7,500 RAF entrants.

IF YOU COULDN'T UNDERSTAND ANYONE, THEY WERE DUTCH

'In 1941, to test security, three men were dressed

up at RAF Codsal in the uniforms of the Kriegsmarine, Wehrmacht, and Luftwaffe and deposited in the middle of Wolverhampton. The man in Wehrmacht uniform was rumbled immediately because of his green uniform. Service police, civilian police and the Home Guard, however, all missed the other two, even though they had swastikas all over themselves! There were a lot of Princess Irene's Dutch Army soldiers in Wolverhampton at the time with orange collar tabs (the Luftwaffe had red tabs, or black, for engineering, yellow for general duties, etc, etc . . .) Eventually, the "Kriegsmarine" and the "Luftwaffe airman" had to give themselves up!'

Len Bradfield, 49 Squadron Lancaster air bomber.

DUSK AND DAWN

'On 5 May 1942 I became a mid-upper gunner in 320 (Dutch) Squadron on Lockheed Hudsons at Bircham Newton, Norfolk. The Hudson was terrible. It couldn't stick anything. No self-sealing tanks. It wasn't made for anti-shipping strikes we flew, during the dusk and the dawn. Thank God we were never attacked by fighters: we would have stood no chance. Altogether, I flew 26 operations on Hudsons, all flown with an all-Dutch crew, mostly at night. (I later flew 75 more operations, on Mitchells, a much better aircraft). My pilot was Ovl.2 (Officer Flyer 2nd Class) Tys de Groot, a bloody good pilot who I loved to fly with. Observer was Ltz.3 Petri. I was a Matr.2 (Seaman 2nd Class) and our W/Op was Anthonie.

'Normally, we took off at around 8 o'clock and returned after three hours. We usually flew at 4,000 ft over the North Sea, sometimes together with four or five others. Soon after taking off you were on your "Jack Jones" (it was too dark to see one another). Losses were high because we flew at low level, belly on the water, straight at ships. Our targets were German convoys off Norway, France, Belgium, and Holland. We'd take off one after the other loaded with four 250 lb bombs. We'd go for two lines of ships, left and right, straight and level. Usually, we approached from their stern. I never fired at ships. I used to save my ammo in case we were attacked by fighters. It wasn't necessary to rake a ship.

'On 27 November, my 14th shipping strike, we started out as dusk was falling in Hudson "L". It was a beautiful evening. There was a calm sea and not a cloud in the sky. We went across the North Sea almost as sea level, and by the time we got to the other side it was practically dark. We were after a convoy off the Dutch coast. By the time we got to Ijmuiden, though, they were gone. We headed north for our secondary target at Den Helder, where my mother and father, a retired Dutch naval officer, lived. They did not know I was flying in the RAF. I had last seen them just before the German invasion of my country in May 1940, after which eight of us, all naval ratings, walked to Dunkirk and finally escaped to England from Boulogne. Some of us now flew under different names so that our families in Holland would not be persecuted. I hated the bloody Germans. I'd seen them bombing Rotterdam and in the *Freedom Paper* it told of how men were put up against a wall and shot and how many boys were sent to Germany. I thought: "Christ! I'm near mum and dad and they don't even know I'm flying!" (Later, in 1943, the Red Cross sent me a letter from them. You could answer them in no more than 25 words. I wanted to let my dad know I was now a sergeant. He being ex-Navy would know that if I said I was now a member of the "Golden Ball" he would know I was a sergeant. (Up to the rank of corporal, uniform flashes were yellow. Sergeant and above were orange). That would have made him proud. My letter survived the German sensor and he knew what I meant. He knew the quickest way to promotion was to fly, and so he knew I was flying).

'We flew up parallel with the Dutch coast. Up front, the W/Op, Anthonie, who came from the Dutch East Indies, was operating the ASV set, scanning the screen for ships. We found two lines of cargo ships 9° north-west of Den Helder, steaming out to sea. Petri, the observer, looked out of the window and saw the ships. A long line of them, steaming slowly eastwards. He was so excited that he yelled out at the top of his voice: "Ships on starboard! Ships on starboard!" He had hardly stopped shouting when they began firing at us. Our formation immediately broke up and we circled the convoy with our bomb doors open, ready to attack. The flak was getting heavy, about eight ships were firing heavy machine-guns and pom-poms and an escorting destroyer was blazing away with a whole ack-ack battery. From every side Petri saw red, white and green tracer fly past, and as we dived for the attack the shells came uncomfortably close. I'd never seen so many golden balls coming up at us.

'De Groot singled out a heavily-laden ship of the convoy. We were going straight, then suddenly, nose down! De Groot opened fire with the two nose guns, giving it everything we'd got, as we dived! I was swearing because we had never done this type of attack before! We'd always gone in straight and level. He kept on shooting and shooting. I thought he would have to stop soon before the barrels melted. Petri stood in the astrodome and warned de Groot where the heaviest flak was coming from and kept a look out for night fighters.

'As we passed over the ship de Groot released the bombs. As soon as the bombs were away, de Groot banked sharply. I don't know if we were hit. (I never ever checked to see if we were, it was always straight into debriefing and straight into the bar for me on return to England). Up front it was a different story. Apparently, a few seconds after bombs away they felt a giant explosion. Petri suddenly heard a loud bang. The cabin filled with smoke and the wireless operator cried out: "I'm hit! I'm hit!" A cannon shell had exploded just behind his back.

'Petri went to his assistance, and as he turned around and looked backwards, saw that the ship we had bombed was on fire. Two of our bombs hit the ship. An explosion sent up a column of flames and smoke a hundred feet high. As we set course for home, the other aircraft [the Flying Dutchmen and the Canadian Demon Squadron, also flying Hudsons] attacked and Petri saw bombs bursting an tracer crisscrossing like "fireworks". Our job was done. We didn't hang around. Petri managed to bandage up the W/Op's wounds. Bernet had several splinters in his back. Although he was in great pain he carried on throughout the return trip sending out messages as if nothing had happened. It was confirmed later that the vessel, a 2,500-ton cargo ship, was sunk. Marvellous! Petri said the sight of all our aircraft, Canadian and Dutch, making a mess of that convoy was worth the many weary months of training and uneventful patrols, and we all felt we had done something in smashing the Nazi's and to repay them for what they had done to Britain and Holland.'

22-year-old Matr.2 (Seaman 2nd Class) Jaap Lub, gunner, 320 (Dutch) Squadron, 1942.

'The Hudson was terrible. It couldn't stick anything. No self-sealing tanks. It wasn't made for anti-shipping strikes we flew, during the dusk and the dawn.' (Lockheed)

HEAVIES

We are the heavy bombers, we try to do our bit,
We fly through concentrations of flak with sky all lit,
And when we drop our cargoes, we do not give a damn,
The eggs may miss the goods yard, but they muck up poor old Hamm.

'The Heavy Bombers'. Airmen's Song Book, *edited by C.H. Ward-Jackson & Leighton Lucas (William Blackwood & Sons, 1967).*

The first of the RAF's four-engine bombers, the Short Stirling, entered service on 10/11 February 1941. In March 1941 it was joined by the Handley Page Halifax. Exactly a year later, these were joined by the Avro Lancaster. (Via Tom Cushing and BAe Manchester)

The first of the RAF's four-engined bombers, the Short Stirling, entered service on 10/11 February 1941. In March 1941 it was joined by the Handley Page Halifax. Exactly a year later these were joined by the Avro Lancaster. By mid-1943 the four-engined 'heavies', using rudimentary forms of electronic warfare equipment, were beginning to achieve a greater measure of success than hitherto possible. The last two years of the war saw Bomber Command grow into an immensely powerful force of some 2,000 bombers. From 1942 onwards, bombing 'round-the-clock' with the USAAF, Bomber Command took the war to the heart of Germany, causing vast destruction to the German war effort. By May 1945 Bomber Command consisted of seven groups controlling 78 operational squadrons of Lancasters, Halifaxes and Mosquitos.

Lie in the Dark and Listen

Lie in the dark and listen,
It's clear tonight so they're flying high,
Hundreds of them, thousands perhaps,
Riding the icy, moonlit sky,
Men, machinery, bombs and maps,
Altimeters and guns and charts,
Coffee, sandwiches, fleece lined boots,
Bones and muscles, minds and hearts.
English saplings with English roots
Deep in the earth they've left below,
Lie in the dark and let them go;
Lie in the dark and listen.

Lie in the dark and listen
They're going over in waves and waves
High above villages, hills and streams
Country churches and little graves
And little citizens' worried dreams;
Very soon they'll have reached the sea
And far below them will lie the bays
And cliffs and sands where they used to be
Taken for summer holidays,
Lie in the dark and let them go;
Theirs is a world we'll never know,
Lie in the dark and listen.

Lie in the dark and listen.
City magnates and steel contractors
Factory workers and politicians,
Soft hysterical little actors,
Ballet dancers, reserved musicians,
Safe in your warm civilian beds,

Count your profits and count your sheep
Life is passing above your heads,
Just turn over and try to sleep.
Lie in the dark and let them go;
There's one debt you'll forever owe.
Lie in the dark and listen.

Noël Coward

'We were to scatter-bomb Mannheim in a "town blitz". Aircraft were despatched to the target singly. At this time we could not afford crew losses from collision or bombs falling on each other's aircraft, so this was the only practical way. It was cold and lonely in the front turret but my Brownings never froze up on any of the operations I would fly. However, if one carried an apple, we were warned not to eat it; it would break your teeth!

'From a safety point of view, the front turret was preferable to the rear, which could be an exhausting trek to reach in our flight gear. The rear gunner was protected by two slabs of armour plate which could be joined together to protect the rear gunner's chest. However, in the front turret I could only rely on armour plate behind the pilots to stop bullets hitting me from the rear. The primary function of an air gunner at this time was to be a heavy flak spotter for the captain. He told us to keep our eyes peeled and report enemy aircraft and flak gun flashes immediately.

'We dropped our bombs on Mannheim from about 12,000 ft and although it was my first experience of flak, we came through all right. We landed back at Newmarket after being in the air for six hours and ten minutes.'

Sgt Alfred Jenner, Wellington WOP-AG, 99 Squadron at Newmarket Heath, 16/17 December 1940. His first operational trip, Operation Abigail as it was code-named, was in retaliation for the German bombing of Coventry and Southampton. In all, some 134 bombers, including 61 Wellingtons, were despatched; the first 'area' bombing raid on a German industrial target.

'We were to scatter-bomb Mannheim in a "town blitz".' (Dr Colin Dring)

Bring Back My Bomber
One night as I lay on my pillow,
My batman awoke me and said:
"I say there are ships in the Channel,
But there's bags of black cloud overhead."

Chorus: *Bring back, bring back,*
Oh bring back my Bomber to me, to me.
Bring back, bring back,
Oh bring back my bomber to me.

So I climbed in my old heavy bomber
And I took off right dead into wind.
And I searched the whole of the Channel
But not a damned ship could I find.

So I turned round and headed for England,
Just thinking of coffee and bed.
The controller said: "How can you miss them?"
I'll leave you to guess what I said!

'I was becoming desperate to get the first op under my belt, to discover for myself whether I would make it – to find out just how scared one would be. Whilst I knew I had a super crew, I also knew they would be watching me very closely to see if I had any of the faults of their previous skipper. Finally, the day arrived and we set off to bomb Cologne. I was very nervous, not at the prospect of enemy action, but having the CO along as my second pilot! It was in fact quite enjoyable and the ack-ack fire was nothing more than a nuisance. At times the Germans seemed keen on playing possum rather than give away the locality of a town by defending it too vigorously. One felt sorry for the crew of the other aircraft, coned in ten searchlights, weaving this way and that, trying to get away, but we were glad they were taking the attention of the enemy.'
Sergeant Pilot Eric Masters, 99 Squadron Wellington pilot, March 1941.

OPS

A biting wind, a searing frost,
A dome of cloud, a misty moon,
Below, a flaming holocaust,
The engines' heavy droning tune,
The spiteful deadly beads of flak
That weave a pattern on the night,
The searchlight cones across the black,
A piercing, vicious blinding white,
The aircraft jinking as it tries
To keep its course and yet evade
The fighters, swarming thick as flies,
That try to stem this cavalcade,

This endless wave, this marathon
Of vengeful bombers, dim and black,
Their crews, relentless, pressing on
To seek, to find and to attack,
A burning bomber hurtling down,
A blazing pyre across the sky,
The incandescence of a town
Alight with flares, prepared to die,
A voice upon the intercom,
The sudden chattering of a gun,
The fire-burst of a fallen bomb,
Its journey made, its duty done.

Audrey Grealy

'One of the most significant features of the German defence system was a mighty bank of searchlights some 20 miles or so long, stretching from about Emden in the north, to almost the French/German border in the south. Not only were these searchlights used in collaboration with flak batteries, but nightfighters prowled in abundance, waiting for an intruder to be "coned". It must be remembered that in 1941 with full bomb load it took a long time to struggle up to 12,000 ft, which was about the Whitley's ceiling, and at that height the indicated airspeed was in the region of 120 mph and the the searchlight barrier was something to be taken seriously . . . I can recall targeting Frankfurt on one night which meant five hours over enemy territory with five hours' evasive action. Not a particularly memorable operation, but reasonably successful and home to bacon and eggs for breakfast and head down until lunchtime, the rub was there was an early afternoon briefing to do exactly the same thing again that night. I was not very enthusiastic, but having revisited Frankfurt and returned safely, my memory of these two raids on consecutive nights is the shattering physical experience.'
Sgt Basil Craske, Whitley pilot, 10 Squadron, August 1941.

'The squadron padre was a jolly soul, and one day he looked at the beautiful, voluptuous female painted on the side of *V-Virgin* and declared: "She's far too broad in the hips for a virgin!"'
Fred Wingham, pilot of Wellington X "V-Victor" (later rechristened "V-Virgin") in 420 Squadron.

'In 1941 with full bomb load it took a long time to struggle up to 12,000 ft, which was about the Whitley's ceiling, and at that height the IAS was in the region of 120 mph.' (BAe)

Night Bombers

Eastward they climb, black shapes against the grey
Of failing dusk, gone with the nodding day
From English fields.
Not theirs the sudden glow
Of triumph that their fighter brothers know;
Only to fly though the cloud, through storm,
through night,
Unerring, and to keep their purpose bright,
Nor turn until, their dreadful duty done,
Westward they climb to race the awakened sun.

Anon

The Navigator

A lot has been written of pilots and gunners
And similar Lords of the Air,
But no-one has lauded the poor navigator
Who certainly also was there.

He patiently sat in his dark little cabin,
Apart from the rest of the crew,
Their guide and their mentor in reaching their
target
And doing what they had to do.

Alone with his charts and his fine calculations,
Their lives and success in his hands,
His brain and Mercator combining directions
O'er enemy waters and lands.

He could not allow any kind of distraction,
Like flak or an engine cut out,
Or jinking or searchlights or somebody wounded
To cause him a shadow of doubt.

He must have had powers of deep concentration
To press on regardless of this,
And calmly sit planning, in spite of the mayhem,
The route back to England and bliss.

His only advantage above all others,
To keep the adrenalin flowing,
Was due to his maths, his dividers and compass —
At least he knew where he was going!

Audrey Grealy

'I'll never ever forget my introduction to this first operational squadron. Arriving around midday, the officer commanding informs me that I would be on tonight's raid. Who, me? Why, I hadn't even unpacked my kitbag. The Flight's chief navigator suggested that he go in my place and I would watch him go through his routine preparing for the flight. I watched him prepare his flight plan, get the meteorology report, go to the intelligence office for his secret coded information, on rice paper so you could eat it if necessary, and other pertinent things to do before going off into that treacherous-looking sky. The navigator and crew never returned. This was a hell of an introduction, especially when stories were circulating about washing out the remains of a tail-gunner with a hose, there was so little left of him.

'I was on the next night. The target was Hanover. Thirteen months after joining the RCAF I found myself in the briefing room nervously preparing my charts for the raid. I was trying to appear calm and nonchalant, this being my first op and not wanting to appear to be too much of a greenhorn. After the briefing, the navigators gathered around the huge plotting table on the operations room and worked out our DR (dead reckoning) courses to get us to the target, and far more important, home again. Our dead reckoning was based on the predicted winds as supplied by the met section, the airspeed, the groundspeed, and the drift, as well as other information so important in our navigation. Many corrections and adjustments were made during our trip from new information obtained in flight. Our only navigational aids in these early days were from fixes obtained from our wireless operator and good only up to a limited number of miles from the English coast. As a matter of fact, I soon learned to jokingly call my navigation – guestimation!

'So here we were, a crew of five, two pilots, a navigator/bomb aimer (observer), a wireless operator and a tail-gunner. I never used a bomb aimer during my tours. They appeared later on in the war and there weren't always enough to go around. I felt that if I could get us to the target I should have the pleasure of bombing same.

'My navigator's table was behind the pilot's seat in the cockpit. As we neared the target I unplugged my oxygen lead, my intercom, and dragging my parachute with me, made my way to the bombsight in the nose of our flying coffin. It was a long crawl in the darkness, and without oxygen the going was tough. Reaching the bombsight and front gunner compartment, I searched frantically for the oxygen connection to restore my strength. With the aid of a flashlight, partly covered so as not to attract any wandering fighters, I found my connection and began breathing easier. I was now lining up the target with the bombsight as I directed the pilot on our

(Handle with Care, *by D. Wesmacott and R. Anderson)*

THE SEVEN DEADLY SINS OF NAVIGATORS NO 7

Over-familiarity with the route.

bombing run – left left – steady – RIGHT – steady – left left – steady – bombs gone. Our Whitley leapt about 200 ft with the release of tons of high explosives. Now we fly straight and level for 30 seconds, the longest 30 seconds anyone will ever know, so that we can get the required photo of the drop for the Intelligence Officer back at base. Picture taken – let's get the hell out of here.

'Still in a cold sweat with the flak bursting around us and the searchlights trying in vain to catch us, I crawled back to my plotting table. The pilot was still taking evasive action as I gave him the course for home. Those black blobs of smoke surrounding the aircraft were flak, and when you can smell the cordite it meant they were bursting too damn close.

'Arriving back at our base without incident gave me a great feeling of relief and satisfaction. It was hard to believe that I'd been over Germany, but harder to believe that I was back in England. Next came our debriefing by the intelligence officers, accompanied by a cup of coffee laced with overproof rum. I was tired, but happy, after our seven-and-one-half hour trip. I guess I had that blissful Number-one-op-behind-me-look written all over my face. I kept thinking: "I've made it". That first op I'd been dreaming about and working toward for 13 months.'

Pilot Officer J. Ralph Wood, RCAF, Whitley navigator, 102 Squadron, Topcliffe, 25 July 1941.

GONE FOR A BURTON

'. . . Every time we beetle down the runway I'm wondering if we're going to make it back. I guess I've seen too many guys go for a Burton this past year. "Gone for a Burton", meaning, in barrack-room language, "gone for a shit". A Burton being a strong ale which caused one's bowels to move rather freely, necessitating a quick trip to the can."'

Pilot Officer J. Ralph Wood, DFC CD, RCAF, 76 Squadron Halifax navigator, Bremen, 3 June 1942.

"You'll be Sorry!"

'Our losses were running at approximately 5 per cent, so one believed one was living on luck after the 20th trip. One was just as likely to "buy it" on the first as on the 30th.'

Eric Masters, pilot, 99 Squadron, shot down on his 30th and final op, on 7 July 1941, to Cologne.

Last Landing

Oft this earth I leave behind,
And soar God's heavens.
Till sun and stars I find,
And fence the towering clouds
With others of my kind.
Fear not if I should lose my way,
Nor keep sad hearts
For my returning day.
'Tis that I flew the heavens too high
And reached God's guiding hand,
And heard him answer to my cry;
"Your journey's done – now land."

A. Burford Sleep

This poem was written by Burford Sleep two days before he 'flew the heavens too high'. A bomb aimer on Lancasters, he was killed returning from a bombing raid on Cologne. He had handed copies of this poem to each other member of the crew the night of the raid, and he was the only one killed in his aeroplane, although others were injured and wounded. Ironically, his brother, Sqn Ldr R.M. Sleep (109 Squadron), marked the target on Cologne on that particular night.

'On our homeward journey we would get into our Thermos of coffee and sandwiches of spam. Of course, our real treat was the flying breakfast of bacon and eggs back at the base and our discussions of the attack with the other crews on the raid. Bacon and eggs were otherwise scarce as hen's teeth. At the ritual breakfast after every mission there were empty tables – chairs, dishes, and silverware aligned – for the men who weren't coming back. Weldon MacMurray, a school friend of mine from Moncton, stationed at RAF Dishforth, about two miles from Topcliffe as the crow flies, once informed me that Johnny Humphrey bought it. Another time, Graham Roger's number came up. This was followed by the news that Brian Filliter was missing in action. One day at lunch I answered the 'phone in the sergeant's mess, and it was Weldon inquiring about me. He'd heard that I'd bought it the previous night. A few weeks later, friends of Weldon phoned me from Dishforth to say that he had failed to return from a trip. I found out

later that he was a prisoner of war. Boy! Was I getting demoralised! Would I be next?'
Pilot Officer J. Ralph Wood, DFC CD, RCAF, Whitley navigator, Hanover, 25 July 1941.

Thanks to Leslie Irvin

To Hell with submarines
They get blown to smitherines
And keep your bayonet chargin'
It gives little safety margin,
Armoured tanks are very fine
'til some idiot lays a mine
And fighting ships at sea
Would scare the pants off me.
If heroics are the need
The words that one should heed
Are "Be backward coming forward"!
Be a calculating coward
And with methods of the day
Fight the safest way
And this by implication
I assess as aviation
Since Irvin was so cute
To invent his parachute.

Jasper Miles

'On 22 July 1942 we collected a brand new Wellington from Oxford factory. Had two flights checking the instruments, synchronising guns etc, and then a few hours rest and off at about 10 pm to Duisburg in Ruhr. I flew as a tail Charlie (rear gunner) – the first to die, or first to live in a crash.

'With a full bomb load we climbed up to 22,000 ft zig-zag fashion as beacons below were directing the fighters out to us. Over the target there was hell on earth. Heavy raid ack-ack plastered a box area of some 300 ft cube, on the Wimpeys and Lancasters which were caught in a searchlight.

'Just after we dropped the eggs and steadied to take the picture, a blue (master) searchlight focused on us with others coning on us. The pilot dived sharply in a hell of bursting shells which peppered the 'plane all over. One shell hit the 'plane, igniting the port engine. The wireless operator was wounded. He called for his mother. Suddenly, the whole 'plane exploded.

'At the time I had manually turned the turret to the right, shooting at the searchlight, but in the explosion I was thrown out. I floated down on the parachute which I had on my chest.

'A short prayer for a lucky escape, a very sad prayer for those in the three burning remnants on the ground, a near miss of the factory chimmney and a wallop with my feet on the railway line. In the morning near Aachen, where I knew a friend's address, I was caught and beaten by the railway's civilians and eventually brought with some Canadian airmen to Düsseldorf army camp for treatment.

'After Dulag Luft I was sent to Stalag VIII-B in Upper Silesia; home to 30,000 airmen and Army, half of them on working parties all over Germany. Many tunnels were dug in the sandy soil and escapes were made through the barbed wire or from working parties. After Dieppe invasion some 3,000 soldiers were caught and sent to our camp, now renumbered 344, but on Hitler's orders they were chained for one year.'
Joe Fusniak, Wellington gunner.

'As the dark horizon of Germany rapidly climbs higher round you, and *Z-Zebra* drops bumping into low cloud, rage grips you again this time at the thought of six men, six friends they are,

(Handle with Care, *by D. Wesmacott and R. Anderson)*

"For you the war is over.'

riding with you and waiting for you to do something, hoping for the act of wizardry that will pull the rabbit out of the fire. Or is it the hat? You can't think which.

'There's Billy, married, by a few months. You never did meet his wife. Don, due to be married in a fortnight. Joe, long since married and content. The rest of the boys, like yourself, with light-hearted dates for tomorrow night. A bloody fine skipper you turned out to be. Thoughts like these loom rapidly into your consciousness to vanish as quickly, pursued by wishful-thinking calculations of fuel and range.

'Like a stab in the back, the starboard inner engine suddenly screams and spews flame. Don reaches for the feathering and fire buttons. He might just as well have sat back and sung the Lord's Prayer. Faithfully he plays out the little game he was taught but, in the language of the times, you have had it.

'Aching with the sheer muscular effort of holding up the plunging port wing, you feel the elevators tighten as the nose goes down with a lurch.

'Too tired to think, you hear your voice giving the queer little order they taught you one drowsy summer day at the operational training unit in pastoral Oxfordshire; the absurd jingle you had never really thought you would ever use:

'"Abracadabra, jump, jump. Abracadabra, jump, jump."'

Geoff Taylor, RAAF pilot of Z-Zebra, shot down over Hanover, from his book Piece of Cake *(George Mann, 1956).*

KRIEGS-GEFANGENEN

'You think I know f***ing nothing; in fact I know f*** all!"'

German PoW Camp 'goon' (guard).

(Handle with Care, *by D. Wesmacott and R. Anderson*)

'. . . and how is your Wing-Commander's eldest daughter Penelope?'

(Handle with Care, ***by D. Wesmacott and R. Anderson)***

'At Frankfurt the reception camp [for all downed fliers] was designated Dulag Luft. Here there was a proper interrogation . . . [but] apart from the odd fag or two the meeting was not very fruitful for either side. In the end my interrogator gave up by saying that I could not tell him anything he did not already know. He knew I was from 10 Squadron based at Leeming and that I was bombing Cologne. For good measure he knew that our CO had recently changed his car and told me the make of it. He was correct of course, but I told him it was a load of codswallop, or words to that effect.'

Sgt Basil Craske, Whitley pilot, shot down 16 August 1941.

'The extent and the manner of the escape, and to get 52 prisoners out in a single night, was a miraculous achievement. I completed about 25 km of the 1,500 km required. Failure? Yes, if success is only judged as being my return to 10 Squadron to fight again, but no if the great cost to the German war effort, both materially and mortally, in dealing with such a large exodus is taken into consideration. The Germans thought so too, and a stir in high places brought a general from Berlin HQ to Stalag IIIE to investigate and produce a report. It was later ascertained that the tunnel was 227 ft in length and was shored with timber throughout – we used about a thousand bed boards in so doing. It is estimated that we shifted about 60 tons of sand. The Germans were impressed and took many photographs of our work. We never did find out who and how many of our keepers spent the winter on the Russian Front.'

Sgt Basil Craske, PoW, who was one of 52 who made a mass escape from Stalag IIIE, Kirchhain, 11 May 1942. He was sent later to Luft III:

'The difference in approach between the IIIE Army guards and the Luft III Luftwaffe guards who collected us for our relatively comfortable

WARNING WIRES

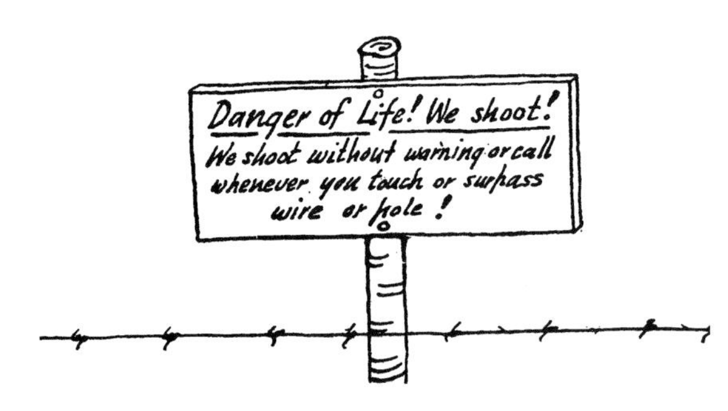

. . . At Heydekrug.

. . . At Thorn.

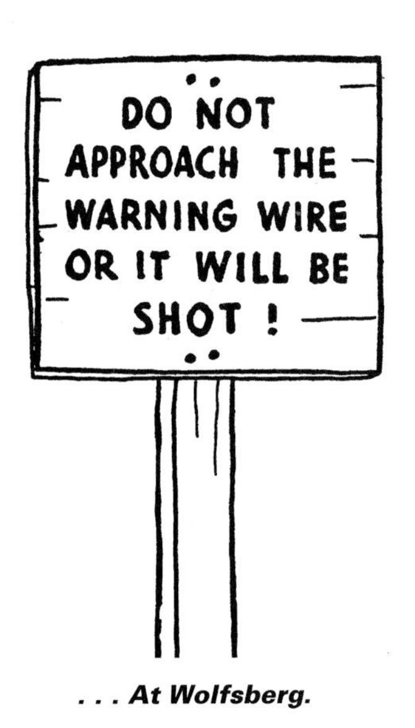

. . . At Wolfsberg.

journey to Sagan, was very apparent. Cool, calm, and 'collected' were the latter, as opposed to the jumpy inefficiency of the Army types. Similarly, the camp at Sagan was in complete contrast to Kirchhain, It was quite vast by IIIE standards and there was room for sporting activities in each compound and even some equipment. Most things were well organised on both sides and our leader, Dixie Deans, must take a great deal of the credit for this. There was a very strong Escape Committee countered by a solid German defence, and by and large one respected the other. For instance, it was difficult for a "goon" or "ferret" to get in (without being tailed whilst in) as it was for us to get out . . . Luft III of course is well known for the documented escapades of "The Wooden Horse" and "The Great Escape" but there were other less serious episodes that raised our morale. For instance, every quarter or so it seemed, a German general would make a tour of inspection and we would gamble upon the colour of the stripe down his breeches. On one occasion the pompous individual ignored the advice of his specialist Kriegie minders and insisted upon showing off his large Mercedes by being driven in it into the heart of the compound. It was of course immediately surrounded by hoards of admiring Kriegies whose one thought was to filch as much of the equipment as possible. Of course there were repercussions, but as a top secret codebook was missing from the glove box we compromised by keeping all the contraband other than their codebook, which after copying was returned to them with every page stamped "GEPRIFT" (censored).

'It was at Luft III that I first observed Sgt Grimson at work, a dedicated and daring escaper who had been shot down in 1939 and unlike the majority of us took the long-term view of the end of the war. He learnt colloquial German by taming a goon in the early stages of his captivity, and thus was very fluent. From an appropriate distance on one occasion I observed Grimson, disguised as a German electrician complete with ladder, walk towards and over the tripwire and approach a guard-tower to explain that there was a fault in the telephone lines which he had to find – they ran along the top of the perimeter fence. By means of accidentally dropping his dummy test meter among the wire between the inner and

(Handle with Care, *by D. Wesmacott and R. Anderson*)

outer fences he managed, with the help of a plank of wood, to get outside to retrieve it. In fact at this stage he was confronted by a guard patrolling the outside of the wire, and by his fluency and his forged pass had the guard resuming his patrol.

'Grimson played a major part in the organisation of escape routes during his many months on the outside of the wire. To achieve this situation there had to be collaboration of a few anti-Nazi Germans and some Germanised Poles, some willingly co-operative and some blackmailed by damaging photographic evidence. Unfortunately, towards the autumn of 1944 the Gestapo caught up with him and he disappeared, as did the German contacts. If ever there was an unsung hero it was Grimson.'

Basil S. Craske, 1942. Stalag Luft III was about 1½ miles from Sagan in Silesia. The camp covered an area of about three-quarters of a square mile.

If you can quit the compound undetected
And clear your tracks nor leave the smallest trace,
And follow out the programme you've selected,
Nor lose your grasp at distance, time and place

If you can walk at night by compass bearing,
or ride upon the rails by night and day,
And temper your illusiveness with daring,
Trusting that sometimes bluff will find a way

If you can swallow sudden and sour frustration,
And gaze unmoved at failure's ugly shape,
Remembering as further inspiration
It was and is your duty to escape

If you can keep the great Gestapo guessing
With explanations only partly true,
And leave them in their heart of hearts confessing
They did not get the whole truth out of you

If you can use your "Cooler" fortnight clearly
For planning methods wiser than before,
And treat your miscalculations merely
As hints let fall by fate to teach you more

If you scheme on patience and precision
It was not in a day they builded Rome,
And make escape your sole ambition
The next time you attempt it – YOU'LL GET HOME

Flt Lt Edward Gordon Brettell, after a bid to escape had failed.

'. . . Luft III, of course, is well known for the documented escapes of "The Wooden Horse" and "The Great Escape".' (via Sotheby's)

'To me, his adoring younger brother, Edward seemed to be not merely a brilliant pilot, but quite indestructible. He had survived a crash at Brooklands before the war when his car went

over the top of the banking. He had also survived when, separated from his squadron in a dogfight, he was pounced on by upwards of 12 Messerschmitts. He shot one of these down and fought his way back to England, though wounded in the head and in a state of collapse from loss of blood. As to his last mission, it went tragically wrong, and not only for him. He was shot down by anti-aircraft fire and, since the cockpit canopy had jammed, went into the ground at over 200 mph. How he was not killed instantly I will never know, although he was severely injured. He recovered from his injuries remarkably quickly and was sent to Stalag Luft III.

'He escaped certainly two or three times. Even then the Germans were thinking of making an example. He and a friend were travelling to Munich by train, as French workers. Unfortunately the RAF began a raid and the train was stopped and then searched. The two of them were arrested as suspected French agents. They had to admit who they were and they were taken before a Luftwaffe officer. He asked them why they wanted to escape. They replied that they considered it their duty to do so. He banged the table and said: "Yes, It is your duty to do so!" After a time and some 'phoning, he told them: "Gentlemen, you are lucky. I am to return you to your camp." It sounds like the Luftwaffe officer had orders to hand them over to the SS or Gestapo, and it also sounds as though, perhaps, he managed to talk someone out of it.'

Terence Brettell.

Edward Brettell was one of 76 prisoners who escaped from the North Compound of Stalag Luft III on the night of 24/25 March 1944 before 'Harry' (the name of the tunnel) was discovered. ('Tom' had been discovered in the summer of 1943 and was blown up by the Germans. 'Dick' was used subsequently to store tools and equipment for Harry.) Fifty of the escapers who were captured, including Sqn Ldr Roger Bushell SAAF, who as 'Big X' organised

(Handle with Care, *by D. Wesmacott and R. Anderson*)

'Wait till he looks the other way, then you hop over the wire."

"I'm sorry, I can't understand a word you say."

(Handle with Care, *by D. Wesmacott and R. Anderson*)

the successful escape, and Brettell, who was caught together with Flt Lts R. Marcinus, H.A. Pickard, and G.W. Walenn near Danzig, were taken to remote spots and shot in the back of the head by the Gestapo. Only Bram van der Stok, Royal Netherlands Navy; Flt Lt Jens Einar Mueller, Royal Norwegian Air Force, and Flt Lt Peter Rockland, RAF, made 'home runs'.

'We Australians, particularly, have no illusions about just what kind of a time our fellow-countrymen are having in the hands of the Japanese. Neither have we any illusions about the misfortune of missing friends who were hanged from German lamp-posts in their target cities or thrown back, maimed and wounded, into the burning wreckage of their Lancasters and Halifaxes by frightened, berserk German crowds . . .'
Piece of Cake *by Geoff Taylor, RAAF, PoW.*

'I remember Dixie Deans, our camp leader, saying at a parade in December 1944: "I've been here for six Christmases and it looks as if I'm going to be here for another bugger."'
Geoff Parnell, PoW.

'The RAF were kept in the special "escape-proof" camp at Heydekrug in East Prussia. This was Stalg Luft VI; and also at Luft I, Barth. Heydekrug was the camp especially built for NCO prisoners. If promotion came through during internment you were promoted from a Stalag to an Oflag. The Germans were very punctilious about this sort of thing; they gave me a receipt for a fountain pen!

'Heydekrug was supposed to be more escape-proof and nasty than Colditz, but 16 NCOs made "home runs" as against 14 officers in the whole of the Second World War.

'I built a mental bridge that made no sense whatsoever. I told myself that the war would be over in nine months. There was no logical reason for thinking this in September 1943. The

"GOONS up!"

(Both Handle with Care, *by D. Wesmacott and R. Anderson)*

"Can you tell me the way to the Sergeants' Mess?'

invasion was almost unheard of. No-one spoke about it or thought about it. We saw ourselves bombing Germany into submission. We knew nothing of the paratroops or gliders being prepared in readiness, although there were books published in 1943 that actually mentioned these preparations!'
Geoff Parnell, air-gunner PoW, Heydekrug, 1943–45.

'Potato peelings are saved and boiled up again as soup for an evening meal. From the two slices of black bread which, with half-a-dozen rotting potatos and a mug of turnip or millet soup, is your ration each for 24 hrs of sub-zero cold, you cut the crusts and shred them into crumbs . . . Mixed with water and a hoarded spoonful of ersatz German jam made from turnips, the crumb pudding is a weekly treat. Generally we have it on Sunday night. It's very important that the pudding should be eaten at night, for it's becoming increasingly difficult to sleep at nights . . . There are of course, fleas and bed-bugs and lice, but then they have always been with us. Bouts of dysentery on a diet of rotting potatoes also disturb the night with hasty visits to the stinking, frozen latrine pit inside the barrack door.'
Piece of Cake *by Geoff Taylor, RAAF, PoW, Stalag IVb, Muhlberg-on-Elbe.*

'Gross Tychow certainly was the toughest camp we had experienced. Nevertheless, we endeavoured as far as possible to maintain a normal Kriegie life with indoor and outdoor activities. In the barracks the usual card schools continued. The cards were not so pristine by now

(Handle with Care, *by D. Wesmacott and R. Anderson*)

"You're CHEATING!"

"I said, have you seen my soap???"

(Handle with Care, ***by D. Wesmacott and R. Anderson)***

"I just want to be alone."

(Handle with Care, ***by D. Wesmacott and R. Anderson)***

and the stakes were lower. The usual clubs had re-formed to discuss matters of mutual interest. The Debaters debated and men of common countries, counties, and districts talked of things still remembered. The Insurance institute we had formed at Hydekrug was not strong enough to survive at Tychow: The study of Life Expectancy tables was hardly an exciting or appropriate subject in the circumstances.'

Basil Craske, Stalag Luft IV, Gross Tychow, Poland, July 1944–February 1945.

SHAKY DO'S

The Twitch

There are many different grades of fear
From simple fright to scared-severe
And each can cause the bowels to itch
This, vulgar airman call "Ring-Twitch".
The twitch, too, has its much or slight
According to degree of fright;
You really know it's dicey flying
If your ring's like an egg that's frying.

Jasper Miles

'This trip was no better than the rest. You'd think by now we'd be used to it. We were all getting the "twitch" as we experienced one "shaky do" after another. A rough translation of "shaky do" is "a very frightening affair". I must admit I was absolutely petrified on many occasions. You had to live with it, control it, but I was lucky. Once the danger was over I got over it fairly fast, until the next op. One of our Hallybags crashed in landing back at the base. The crew and aircraft were a mixture of broken bodies and metal. My morale was sure taking a beating.'

Pilot Officer J. Ralph Wood, DFC, CD, RCAF, 76 Squadron Halifax navigator, 19 June 1942 (his 22nd op).

"Fairly shaky do."

(Punch)

'On a cross-country practice trip we lost an engine (the engine had seized). The machine-gun-like noise and vibrations in the aircraft startled the hell out of us. It was so scary to our sensitive nerves that we nearly bailed out. As a matter of fact, the flight engineer had already kicked out the door of the escape hatch in my compartment and away it went toward the English countryside. Who said we weren't nervous.'

Pilot Officer J. Ralph Wood, DFC, CD, RCAF, 76 Squadron Halifax navigator, Emden, 20 June 1942 (his 23rd trip).

'Op No. 24, Bremen, 25 June 1942.

'It was a pretty spectacular raid, overshadowed by the incident I was to witness on our return to base. Another total loss of an aircraft and its crew. They crashed and burned on landing. I'll never forget the spectacle of bodies trapped in the aircraft, the reek of smoke and the fumes of death. Not a pretty sight! It was all I could do to keep from throwing up. It's no wonder my nerves are wearing pretty thin. Again the word "sabotage" was on everyone's lips. The rum and coffee was sure needed at interrogation this night.

'On the following afternoon, 28 June '42, our crew were detailed to do a cross-country exercise over the southern part of England. The purpose of this exercise was to give a newly arrived navigator to our squadron some more practice. I went along in case he ran into difficulties with his navigation. We also took along one of our ground-crew men for the ride. Halfway through the exercise, as I was lying on the bench in the crash or rest position, midway in the aircraft, a sudden loud machine-gun-like noise was heard. This was accompanied by tremors and vibrations throughout our Hallybag and scared the living hell out of us. One of our starboard engines had seized and Tackley had to feather the blade in order to regain control of the aircraft. We were later to learn that oil leads to the engines in the Halifax would sometimes break due to the vibration and cause the engine to seize.

'As we continued on our journey and were nearing our base, the tail gunner asked me to sit in his turret for the landing. It was required to have someone in the tail for proper distribution of weight when landing and taking off. I declined, so he asked our ground-crew passenger to go back in the turret. While he went back to the tail, I took up my favourite position when landing in daylight. This was in the mid-upper turret with my heels on the ladder facing the front of the aircraft. This gave one a great view of what was going on, through the blister. We were about 1,000 ft off the runway when the other starboard engine seized, making that same terrible noise. With two engines on one side and none on the other we swung sharply to the right, lunging toward a farmer's field. Tackley never had a chance to feather the blades of the second engine. Viewing through the blister what looked like impending disaster, I frantically dropped to the floor and hung on to the bottom of the mid-upper turret ladder. I kept thinking: "I'm going to look pretty foolish if nothing happens". The next thing I knew I was bouncing back and forth together with the target flares, which had become dislodged from their usual position on the walls of the aircraft. When everything stopped rocking I immediately made for the door at the rear. After three recent fatal accidents at our base, I was more than determined that I wasn't going to burn. Reaching the rear door in what must have been record time, I found it to be jammed. I turned in a panic to the wall opposite the door, where an axe was usually kept. Maybe I could chop the door open! As I turned I saw daylight where the aircraft had broken apart, and quickly made my exit. My legs were so shaky I fell to my knees.

'As I looked around I noticed Tackley on his back with his clothing smouldering lightly. I rubbed out the embers, thinking how fortunate it was that we didn't have a full fuel load at this time. I had no idea whether he was alive or not. I later learned that he hadn't made it. The flight engineer and wireless operator were both pretty gouged, but eventually recovered. The groundcrew laddie, who filled in for the tail gunner, was found in the next field. The aircraft first on one wing had slung him out of the turret like a clay pigeon from a catapult. His injuries consisted of a broken neck, a broken leg, a broken arm, and a great many bruises. When I visited him in the hospital later he told me he was recovering quite well but didn't think he'd do any more flying. Apart from a few bruises and one hell of a scare, I was uninjured. Our Hallybag was literally in pieces. The fuselage was broken into at least two sections. The wings were torn off and the engines distributed around us. The meat wagon (ambulance) and fire engines from our base were soon tearing across the fields and took us to the station hospital. We were given a large belt of overproof rum and a quick medical, then dismissed. That is, those of us who were still mobile.

'I thought, as we made our way to the mess for tea: "This has got to be a dream". I shook for two hours as this dream became startling reality. Today I came damn close to cashing in. Twenty-four ops and we had to crap out on a lousy training trip. That night I went to the dance at Darlington where Tackley had a date with his girlfriend. I had to give her the bad news. I never heard a girl scream before. As soon as possible I returned to my barracks and tried to sleep. It was an experience I will never forget, ever.

'These four crashes around our airfield, ours being the fourth, were playing havoc with my nervous system. I requested sick leave from the squadron medical officer, who was reluctant to recommend same. Instead he sent me to a Canadian medical centre in East Anglia, where I found the reception filled with other nervous aircrew, most of them there to go LMF (Lack of Moral Fibre). This meant that they would be stripped of all rank and placed on general duties. They would be in disgrace, but they would remain alive. I thought to myself: "There but for the grace of God go I". The most unlikely people, people like myself, could usually rise to the occasion, just as big, tough guys fold under pressure. You never knew. I was interviewed by a Jewish Canadian medical officer whose first question was: "Do you want to quit flying?"

'I think he was a little surprised when I assured him that all I wanted was a little time off to get away and put things together again before continuing my tour.

'Armed with his recommendation for two weeks' sick leave, I returned to our base, packed my kitbag and headed for Dunoon, Scotland . . .

'Our Hallybag was literally in pieces. The fuselage was broken into at least two sections. The wings were torn off and the engines distributed around us.' (via Steve Adams)

I also visited Edinburgh during this leave. It was here that a new experience was added to my assortment of nightmares. I'd wake up in the middle of the night to find myself standing on my bed and trying to push my way through the wall. I presume I must have been trying desperately to get out of an aircraft. Another version of this antic was a leap out of bed and finding myself standing beside it, wondering what the hell was going on. It was several years after the war before I was able to rid myself of that annoying performance.

'Op No. 27. Bremen, 4 September 1942.

'It turned out to be a hazardous experience, magnified by my state of nervousness. On our return to base I said: "This is it", and approached the commanding officer, who agreed to retire me from ops, crediting me with a full tour. In the heavies, a tour consisted of anywhere from 25 to 30 ops. I had 27 to my credit.'

Pilot Officer J. Ralph Wood, DFC, CD, Halifax navigator, 76 Squadron. (Altogether, Ralph Wood completed two tours (77 ops) by November 1944.)

'The aircraft hit with a hell of a crash. The props flew off, engines ran away and quickly caught fire. We bounced on some high ground, became airborne again. The next contact with the ground was a rail embankment, into which we nose-dived. We had 1,500 gallons of high-octane fuel aboard. The fuel tanks exploded on impact and, lit by the engines, became a wall of flaming petrol. The aircraft broke its back just behind the second spar, flipped over the embankment and Sgt Michael Read, the bomb aimer, and myself, were airborne for about 70 yards. We passed through the wall of petrol and became flaming torches, landing in a foot of mud, water and lots of bullrushes, where we burned quite furiously. Fortunately an old carpenter working nearby ran to the crash as fast as his 72 years would allow. He looked at the carnage, decided there was no hope of survivors, and came over to investigate the two columns of smoke some distance away. He found us and rolled us into the mud to put out our fires. Apparently my uniform by now was almost completely burnt off. Read was in slightly better shape, although his arm was broken in two places. I had lost my helmet and was fairly well singed about my face and hands. My shoulder blade was broken and about six ribs were broken. A sizeable lump of flesh was grounched out of my groin and a perfect print

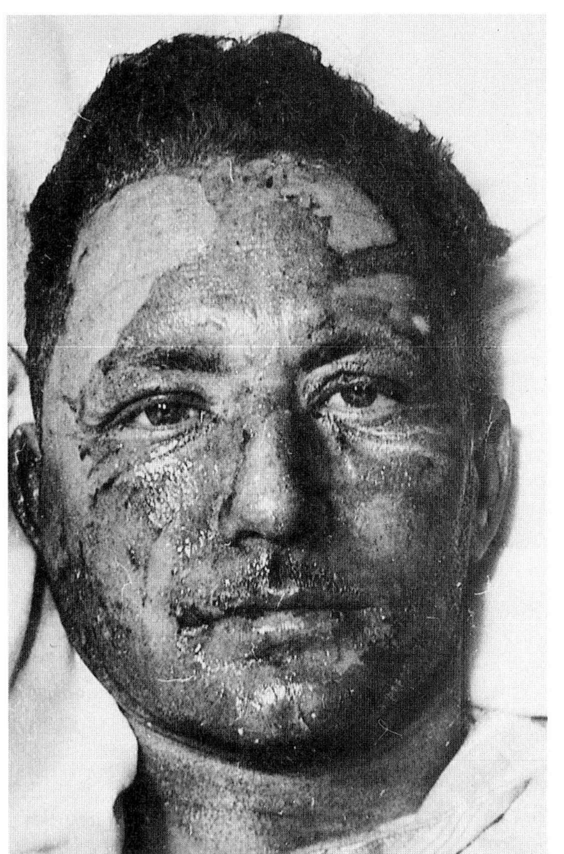

'I became one of Sir Archibald McIndoe's "guinea pigs" at East Grinstead hospital. I was photographed every morning and got a copy of my early grafts.' (Flt Sgt Ivan Williamson via Tom Cushing)

of a Lancaster spar, including two bolt heads, was imprinted on my back.

'It seems incredible that a Red Cross nurse should be waiting for a train half a mile away with her emergency kit. She was soon on the spot and administered injections. I was apparently quite violent, insisting she shouldn't waste her bloody time on me, but concentrate her efforts on the rest of the crew, who were much worse off.

'There seems no doubt that we would have burned to death if the carpenter had not been there. He had picked me by my "Kiwi" accent. I insisted I was out cold, but he said: "Oh no boy, your language, it were real thick. It were real bad." Most of it was directed at the nurse for wasting her time on me. I hung my head, a little ashamed, but I had no idea about all this.

'We were taken to Ely hospital and I was an in-patient for two months. Officers of the investigating committee questioned me.

(via Tom Cushing)

Apparently the only part of the aircraft completely whole was the engineer's panel.

'I became one of Sir Archibald McIndoe's (a fellow New Zealander and father of skin grafts) "guinea pigs" at East Grinstead hospital. I was photographed every morning and got a copy of my early grafts. I healed like a healthy animal and it wasn't long before I was disgracing myself, getting a hard-on in my saline bath.

'Life was cheap those pretty hard days. I went back to the squadron as a lost soul and crewed up twice with new crews but was taken off the battle order by Wg Cdr Anmaud, a great airman, both times. Neither crew survived their first op. I later crewed up with a crew who had seven trips in Stirlings, but the pilot and bomb aimer went LMF. They were stripped of all rank and the crew were finally broken up after six trips, four of which were to Berlin. I went on with 75 (New Zealand) Squadron later and in all did 44 operations. I had incredible luck all the way through. It would be no exaggeration to say I have missed death at least ten times during war and since.'

Flt Sgt Ivan Williamson, DFC, RNZAF, W/T operator, 115 Squadron. His Lancaster Mk II crashed at Magdalan, about 4 miles north of RAF Downham Market, after feathering two engines at 1,200 ft during an air test on 14 September 1943. Flight Sergeant Bert Bradford, the pilot, and the rest of the crew were killed.

'He sat on the steps of the billet, his arms and hands moving through the air in a scatter motion as he fed the "chickens"; except, of course, there were no chickens. They took him away a few days later.'

Lancaster pilot.

MALTESE TIMES

'In the future, after this war, when the name of Malta is mentioned, you will be able to say with pride: "I was there".'

AVM Sir Hugh Pughe Lloyd, 20 April 1942.

Throughout the Second World War the RAF's commitments overseas were considerable. In the Middle East and North Africa, after initial setbacks, the RAF played a major part in defeating the Axis forces. The RAF fought alongside units of the USAAF under the control of the combined Mediterranean Allied Air Forces and eventually swept through Italy, the Balkans and southern France. Having begun the war with meagre forces based mainly in Egypt, the RAF had come to dominate the region by 1945, with 67 squadrons operating from major bases in Italy, North Africa, Greece, Malta, Egypt and Palestine.

'On Tuesday 14 April 1942 148 Squadron changed aircraft from the Merlin engined Wellington to the Pegasus radial engined Wimpy. The next morning I was pulled from my bed at 05:00 hrs by the squadron leader. "Come on, wakey wakey, we're flying in half an hour. Get cracking!"

'"What the hell is going on?", I thought. I still had a "thick head" from last night. I dipped my head into a pail of water, pulled my shorts and shirt on, and made my way to the dispersal where the aircraft was waiting. The engines were ticking over, and as soon as I was aboard we took off.

'As soon as we were airborne I asked: "What in hell is this all about?" "We are on altitude test with full bomb load to see how high the Wimpy will fly on the Pegasus engines, that is all." We climbed to 15,000 ft in just over half an hour. When we landed we were told: "OK chaps, you can go back to bed now."

'We still did not know what the exercise was in aid of. At the same time the next day we were on it again. Again we reached 15,000 ft, but could not get above that. This was the highest I had flown in the desert. Our normal height on the operations was about 10,000 ft, if we were lucky.

'Two days later we were told what the 'big job' was. AVM Hugh Pughe Lloyd had ordered that eight Wellington bombers be sent to Malta to attack the airfields in Sicily, where the Germans operated from to attack Malta. F*** my luck. I was picked to go.

'I was the rear gunner in the Wing Commander's crew. Each aircraft would carry a spare crew. Jim [Crank-Benson] was a passenger in our kite and Brownie was the wireless operator. No big beer-up in the mess tonight. Everyone to bed early.

'On Monday 20 April we took off from Kabrit for Advanced Landing Ground [ALG] 106. Eight Wimpys, 16 crews and 96 aircrew and a few ground crews. We flew up to the ALG in formation as there had been an increase in fighter activity in the area lately. After a flight of 1 hr 35 min we were on the circuit at 106. There was a sandstorm blowing, making the landings a bit dodgy. It was our turn to go in. Visibility was poor and the approach was bumpy, and just before touchdown we stalled, hitting the deck with one almighty crash. Although the aircraft was "bent up" a bit there were no injuries.

'Sgt Vinall, who was our pilot from base, was replaced. He had to wait for a spare aircraft and join us later at Malta. The gear had to be loaded into the spare aircraft that was at the ALG. Engines were run and other checks were made. Then it was time for the meal. Take-off was 20:00 hrs, so I had time to write a few letters home.

'Time for take-off, engines were started, the final checks were made. We were on our way. After an uneventful flight of 6 hrs 20 min we were on the circuit at Luqa. The runway was well lit up. There were no air raids on, and on landing we were told to get our fingers out and get over to the dispersal by following the truck in front. Once again I was on Saffi strip (I had been there in October 1941). I thought that I would never see this again.

'When we reached the dispersal and engines were switched off I was the first out, and I asked the truck driver what all the panic was. He said: "You will see in a few hours' time". After the debriefing at the ops room we were told that we would be staying at the palace at Naxxar (Nasher), and to stay out of sight when the raids were on. Jerry would not bomb the towns unless they were occupied by troops. That is why we were housed there. When we arrived at the palace it was hard to believe that this would be where we would live for our stay on Malta. About 30 wide stone steps led up to the entrance. (I was expecting a footman to open the polished doors for us.) Inside the large hall were paintings on the walls, and a well polished floor led into rooms on each side. Most of the furniture had been stacked at one end of the room, and this was replaced by RAF beds. What a contrast to the hut in the quarry where we had stayed in October 1941.

'On 20 April 47 Spitfires were flown off the USS aircraft carrier *Wasp*. Within 90 minutes after the first one had touched down, Jerry had made yet another raid on the airfield. Seventeen Spitfires were destroyed, 23 were damaged and many had been lost in combat. Later in the month the enemy bombers attacked stores,

'I woke up next morning to the sound of air raid sirens. Not one of us bothered to get up. Then we heard the guns opening up, and, although the bombs were some distance from the palace, the building began to shake. Get to Hell out of here.' (RAF Museum)

barracks and camps, and hospitals, although they were clearly marked.

'As soon as we got sorted out it was time for kip. We had been on the go for 22 hours and it didn't take any time to get to sleep. I woke up next morning to the sound of air raid sirens. Not one of us bothered to get up. Then we heard the guns opening up and, although the bombs were some distance from the palace, the building began to shake. Get to Hell out of here. There were slit trenches at the rear of the building, so we made a dive for them. Naxxar was on high ground, so we had a good view of the bombing.

'The yellow scramble flare was fired from Luqa, warning that a raid was imminent, and to get fighters airborne. Children were pointing out to sea. They had seen the first wave of bombers approaching the island. Then I could see them, about 50 in the first wave. As they got close the guns began to open up. Soon the sky was full with black puffs of smoke and I could recognise the aircraft as Ju 88s. From about 15,000 ft they started to dive on their target. This time it was Grand Harbour. It was the turn of the light anti-aircraft guns to open up. I could see the tracers snaking up at the bombers as they dived. "There go the bombs!", somebody yelled. A huge pall of black smoke was hiding the harbour from our view.

'As the first wave pulled away, another wave was coming in. This time it was the Ju 87s, the Stuka dive-bombers. Their target was the airfield. As they dived, almost vertical, I could see the bombs leaving the aircraft, and at the same time they would pull out of the dive and head for the sea. I was watching one aircraft. He had released one large bomb . . . Christ, is he going to pull out? No, he didn't. As he crashed into the deck with an almighty explosion I could hear the islanders cheering. They did this each time they saw a Jerry shot down.

'Our fighters were in amongst the bombers and the ack-ack. They were outnumbered by six to one. At the time we only had half a dozen that were serviceable. I saw one fighter coming in to land, probably out of ammo or fuel, and as he was near to touchdown two Bf 109s attacked him. The fighter burst into flames. The pilot had no chance of getting out.

'As more bombers came in, the sky was full with shell bursts, the black smoke from each shell forming into one big cloud. Smoke and dust was rising over the island. The sound of the guns and bombs was deafening. Amid all the noise I could hear the islanders cheering as enemy aircraft were brought down.

'The last wave had finished bombing and were heading out to sea, on course for their airfields in Sicily; airfields that we would be bombing later. Quiet reigned over the island. Smoke and dust was still rising and fires were burning on the airfields. Way out to sea a seaplane was picking up the airmen who had baled out or had ditched in the sea.

'Three more raids were to follow that day, including the Italian raid. Six Italian aircraft, flying in vic formation with one at the rear at 15,000 ft, released their bombs simultaneously flying on a straight and level course, leaving a trail of flak puffs behind them. They would hold this course across the island, and never did I see one shot down.

'The Italians were the first to bomb Malta in 1940. They failed to make an impression, although they killed and wounded 200 Maltese and destroyed 350 houses. In January 1941 the Germans began to bomb Malta, and in the next 4½ months had killed as many again. Later, the Germans were called away, leaving the Italians to go it alone. During the next seven months they killed a further 100 civilians and destroyed 300 homes.

'The Germans returned, and between 21 December 1941 and May 1942 they killed 800 Maltese, nearly 2,000 were injured and 4,000 buildings were destroyed. Malta had had 2,148 raids up to today, 21 April 1942. Between 24 March and 12 April 2,000 sorties had been made against the Grand Harbour alone. During April, 2,000 sorties (284 raids) were flown against the island. Among the targets were the three airfields of Luqa, Takali, and Hal Far.

'After each raid the bomb craters had to be filled in, runways repaired and cleared of debris. This would occur several times a day. The Royal Navy, soldiers, civilians, and even the Maltese Police, assisted the understaffed RAF ground crews, and at the end of the day they had had a "gut full". Valetta was out of bounds. Two of the boys did break bounds, and got killed in the raids. I thought about the bar we used to go in, the mother, and the daughter who George fancied, and the Under 20 Club. I wondered if they had survived the bombing. I thought of the good times we had, the six of us. Now there were only three. How many more were going to make the boat?

'A lot of people had lost their homes and were

living in caves at the quarry (when we were here in October 1941, we spent one night in the hut at the quarry). I was lucky to be invited into one of these caves. One large tunnel had been cut into the rock, and from this smaller tunnels were cut in the sides. In each of these "rooms" lived one family. The woman who invited me in was holding a lamp made from a tin with a piece of string as the wick. In the dim light I could see bits and pieces of furniture that they had rescued from their bombed home. A small cooking stove, a rickety chair, a very old clock, a deckchair, and boxes of cutlery. A family of eight lived in here. The woman pointed to a small hole at the end of the cave. It was an altar. An oil lamp was burning, and at one side was a small bunch of flowers. On the other side was a picture cut from the *Maltese Times*. It was a picture of a fighter pilot who had been killed on the island. It was here that they prayed. Not for themselves, but for the fighter pilots, the men on the guns, and the sailors on the convoys at sea.

'A few chickens were running about outside, so what with the eggs and an occasional chicken it helped to supplement their small rations of food. Although there was a shortage of food, especially of grain, the people of Malta were bearing up. Bread was rationed to just a few slices per day. The mother asked me if I would like a cup of tea. I said not to bother owing to the strict rationing. She insisted I have one. It was in a glass, no sugar, no milk, but with a slice of lemon in it. That was the first time I had tasted lemon tea. I marvelled at how the people of Malta stood up to these long, hard months of continuous bombardment.

'After each raid the bomb craters had to be filled in, runways repaired and cleared of debris. This would occur several times a day. The Royal Navy, soldiers, civilians, and even the Maltese Police, assisted the understaffed RAF ground crews . . .' (Author's collection)

'At last, on 22 April, we were going to get our revenge on Jerry. Ops were on and we were taken out to the aircraft to make ready for the "big one". Our aircraft had survived the first raid of the day. I was busy in the turret when the yellow flare went up from Luqa. Soon the sirens were wailing and I could see the first wave of bombers out to sea, heading for the island. The kite was in a sand-bagged enclosure. Should I stay with the ol' gal, and get blown up with her, or run to the slit trench and get buried alive? It was quite a long way from Saffi strip to Luqa, where the underground shelter was. As I was making up my mind the first bombs were falling. This wave were bombing the airfield at Takali, which was about five miles away. Hal Far airfield was only two miles away. I thought my luck was in. Just then a lorry appeared from nowhere. "Want a lift to Luqa?", the driver shouted. I didn't stop to get in the cab, just stood on the running board as we sped off to Luqa.

'We made a dash to the underground shelter. Just made it! The next wave was bombing the airfield at Luqa and Saffi strip. Jim was in the shelter, and he told me that I was flying with him tonight. This would be the first time he had flown on ops as captain. On all other ops he was "second Dicky".

'The "all clear" was sounded. It was time to return to the palace for a meal and to get our gear ready for tonight. The take-off would be early as it was only a three-hour trip, and it was planned to get two, or possibly three trips in that night.

'Just before take-off a coach was sent out to take us to the aircraft. We had to run down the steps of the palace and quickly get into the coach. As we were dressed in flying gear, complete with Mae Wests and parachute harness, the waiting crowd outside in the square began to clap and cheer. This move was supposed to be secret, but they knew what we were about to do. A film star would have liked this. In fact, I think they imagined that we were.

'On the way to the airfield another raid had started. We had to scatter. There were no shelters or slit trenches nearby, so we left the coach and took shelter near a stone wall. This raid was on the airfield at Luqa. When the raid was over and we reported to the ops room, we were told that several of the kites had been hit (ours was OK), and it would delay the take-off as the runways had to be repaired. These crews that had lost their kites would have to wait for the ferry crews to come in from Gib.

'We had lost quite a lot of kites through the bombing, and the ferry crews that were bound for Egypt were diverted to Malta so that our squadron could take them on ops that night. Most of the kites were bombed the next day and the same thing would happen again. This was causing a build-up of spare crews and not enough aircraft.

'The briefing was over, a lorry took us out to the kite. A final check by the ground crew for any shrapnel damage, everything seemed OK. The engines were started. We began to taxi up Saffi strip to the airfield, where we waited for the "green". Jim was making a final check on the engines. We got the "green". At last we were on our way.

'Sicily is only 60 miles from Malta, and the target, the airfield at Comiso, was about the same distance inland. As we crossed the coast we were greeted with the first heavy flak. We were flying at 8,000 ft, so most of the shit was bursting above us. We ran into several of these heavy ack-ack sites on the way to the target, and I could see other aircraft bombing airfields as we were heading for our target.

'The navigator called Jim on the intercom. "OK Jim, I can see the target." The searchlights were probing the sky and as we turned I could see flares drifting down, lighting the target up like day. The light flak was spiralling up and I could see bombs exploding across the airfield. There was a mountain range about four miles

'We had lost quite a lot of kites through the bombing.' (RAF Museum)

from the target, and Jim said he was going to go in over the mountains and make a low-level run over the target. The last of the aircraft in front of us had completed their bombing. The searchlights began to fade and the flak had stopped. We had the target to ourselves. Jim carried on. The navigator said: "You'll get a bollocking when we get back!"

'Jim replied: "Goin' in now. Here we go." Over the mountain, a steep dive and we were running up to the airfield. Then the flak and searchlights came to life. The navigator was having a hell of a job to see the target. Then: "Hold it steady Jim, I can see it now. Steady . . . left . . . steady. Bombs gone!"

'I was shitting bricks in the back, and I could see the ack-ack on the side of the mountain was having a go at us. I fired my guns in that direction but I knew it was useless. I think I was firing to take my mind off things down below. We were carrying 40-pounders. The bombs had a "mushroom" on the nose, and as they hit the deck they would explode at ground level, scattering shrapnel across the airfield. We left quite a few aircraft burning. Things were a little quieter now, and I could still see the fires from several miles away. We ran into more heavy and light flak on the way back, but this was nothing compared to the target area. As we crossed the coast Jim handed over control to the second pilot. At this moment there was one hell of an explosion. The kite was tossed on its side. Jim took over control again, and at the same time yelled at the second pilot: "What the hell did you do!?"

'"Not a bloody thing," replied the second dickey. "We've been hit!"

'Jim told me to have a shuftie. I plugged my intercom in.

"What did you find Wal?"

"Just a bloody great hole in the port side," I replied.

'Jim said he would go back to have a look at it. I climbed back into my turret, and plugged my intercom in just in time to hear Jim say: "Jesus bloody Christ!" Was he praying? I don't think so.

'Soon the beacon at Malta was in sight. We got the OK to land. Then began the long taxi to the dispersal at Saffi strip. The engines were switched off. As usual I was the first out to light up that cigarette. Never had I enjoyed a fag like that one. As Jim came out the ground staff bod saw the damage and said to Jim: "How the hell did you get this back?"

"I didn't," said Jim. "The flak blew us back."

'A truck took us over to the ops room for the debriefing, and we were told that there was not another kite for us. We could return to the palace. After a meal I stayed on at the airfield, helping the boys to bomb-up. Most of the crews were getting two trips in. A couple of crews managed three. That was good. A photo reconnaissance aircraft was sent out later, and according to the photographs the boys had done a good job. But Jerry was back again the next day. Nobody expected that we were going to stop the raid, not with eight aircraft! If we had 800 we might have made an impression. We had eight aircraft at the start. Ours was u/s, two were shot down over the target, and one was shot up and damaged. Total aircraft at daybreak: four!

'On Thursday 24 April our kite had been repaired, or patched up. Jim gave it an air test. Just one circuit of the airfield. When we got back to the dispersal I checked the guns and ammo before the first raid of the day started. Today they were early. Just my luck. I was caught on Saffi strip again. "'Ere we go again, what to do now?" It was then that I noticed a trail of dust coming from Hal Far. Wonder if it's coming this way? With a bit of luck the driver saw me. He did. "Want a lift to the shelter?" . . . We only just made Luqa and the underground shelter in time. The siren was wailing and the guns opened up. "150 on the plot" we were told. This was going to be a big one. The shelter and ops room were a good 60 ft below ground. I was sitting near the vent shaft, another bod was sitting opposite. It was then that I witnessed something I had never seen before, or since. Suddenly, his lower jaw dropped open and he was making "queer" sounds. At the same time he was pointing to his mouth. I called for the MO. He told me that his jaw was broken from the blast of a bomb that had exploded near the vent shaft.

'The raids were heavy. It seemed as though Jerry was taking the mickey for what we did last night. I wondered if we would be on again tonight. Was it worth it? We lost several aircraft and crews, for what? Everybody knew that it was fighters, not bombers, that were needed to stem the raids.

'Ops were on again tonight. Our kite was hit on the ground during one of the raids. There were no spare aircraft, so we were to be the "standby" crew in case of sickness by any of the crews. Two kites missing last night and, with Jerry knocking out our aircraft during the day, the squadron was in a bad way. We were told the next day that the squadron was to return to Egypt. I was not sorry to hear this. I had completed 34 ops and the old "ring" was beginning to "twitter", as we used to say.

'At 5 p.m. on Sunday 27 April we left the palace at Naxxar dressed in full flying kit, once again to the cheers of the locals in the square. The *Maltese Times* gave us a big write-up regarding how we had bombed the enemy airfields in Sicily. This had raised the morale no end. As we boarded the coach they were still cheering. Little did they know that we were leaving them for good.

'The Commanders-in-Chief, Middle East Forces, had sent a request to the Chief of Staff in England, requesting that they send heavy bombers out to bomb the port of Tripoli, and Italy. At the end of the month the Governor of Malta made a similar request, but was told that this was out of the question as the heavy bombers were building up in strength to attack Germany from the UK. Everybody, except those at the top, knew Malta wanted fighters, not bombers.

'After the briefing we had to wait in the coach, two crews per coach, for the aircraft to arrive from Gibraltar. Soon, our kite arrived. It was refuelled, and the gear was stowed inside. The ferry crew were not pleased at all that we were taking their aircraft. They would be shipped back to Blighty by sea later.

'Tonight I was flying as passenger. Jim was in the kite also as passenger. Wing Commander Rollingson DFC was captain, with Sqn Ldr Pricket as second pilot. As soon as we were airborne I heaped a pile of flying clothing on one side of the kite and got my head down. We touched down at Kabrit at 08:30 hrs, after an uneventful flight. As we left the aircraft to board the coach we were greeted with: "Cor, you should have been 'ere last night. We got raided by Jerry. One 'anger wos 'it . . ." The cockney driver prattled on and on. "How many kites?", he was asked. "About 'arf a dozen He 111s," he said. We did not tell him that on the last raid on Malta 150 took part. This was the first time Kabrit had been bombed. Rommel was pushing the 8th Army back towards the Canal area, which meant he could bring his bombers down from the desert.

'The month of April 1942 will be remembered by the people of Malta, and those who were stationed there. That was the month when His Majesty the King awarded the George Cross to the people of Malta, the first time the award was made to the people of one country.'

Sergeant Wally Gaul, Wellington rear-gunner, 148 Squadron detachment on Malta, April 1942.

'Battle scarred George Cross Malta stands firm, undaunted and undismayed, waiting for the time when she can call: "Pass friend, all is well in the island fortress!"'

Governor, Lord Gort, Palace Square, Valletta, 13 September 1942.

THE 'MAIL RUN' TO BENGHAZI HARBOUR

Take off for the Western Desert,
Fuka, 60 or 09,
Same old Wimpy, same old target,
Same old aircrew, same old time.

'On minelaying runs to Benghazi we did an afternoon take-off from Shallufa with full weapons plus maximum fuel, no overload tanks, and flew the two-hour flight to an advance desert landing strip, usually LG09, to top-up the tanks. It was always a dodgy landing on near-virgin desert with a Wellington and a large load. 38 Squadron was the only squadron on minelaying duties. When there was a shortage of mines we used tailfin cartons filled with sand. We would make raids with these, interspersed with real mines. Our magnetic mines could be set in any of 16 polarity positions. Enemy shipping movements, therefore, could not take place until their minesweepers had made 16 sweeps over all the approaches to Benghazi. Dropping mines required three or four runs. Minelaying involved two crews working every second night, but it was usually very busy with bombers overhead because the attack would involve other squadrons bombing the town and harbour heavily. All aircraft, therefore, were timed precisely at half a minute each at heights between 10–12,000 ft. Minelayers were to go in at 200 ft during melée to drop close-in, either inside the harbour or up to half a mile on deep approaches.

'We had a meal at LG09 at about 6 o'clock and had a detailed last briefing. Take-off for Benghazi was 22:00 hrs. Engines run up on *"U-Uncle"* revealed port engine gills adjustment were faulty. Switch off. Called engine mechanic. On examination he said he could set gills at an optimum position then he'd have to adjust. The delay meant we would have to "drop" an hour after main attack and without other supporting bombers: ie, alone! Also, we were ordered to drop one mile out from the harbour entrance. It was thought to be less fraught!

'We took off from LG09. Using aircraft letter 'U' we telegraphed in morse "UI, UI, UI". (On the way back we'd telegraph "UO, UO, UO'.) No other transmission as enemy could fix our position. Next aircraft in sequence listening followed on. We crossed the coast from the south. Bright moon to eastward. No cloud. Shore silhouetted. Flew close in under cliffs. On sighting harbour approach, banked out to position one mile out 200 ft altitude. No lights of any description to be seen ashore. Enemy did not suspect a "lone one". We made our drop run north. Could have got away with it but the "Mickey Mouse" selector did not let the mines go. Flight Sergeant Roger Mannel, the pilot, said on intercom: "We'll turn and run back with same track to drop". Once I flew with a captain who made 13 runs because he wasn't satisfied! The crew played absolute hell with him.

'Round we came, levelled; nothing from the shore! The enemy, too late, realised our north run

'Dropping mines required three or four runs. Minelaying involved two crews working every second night . . .' (R.G. Thackeray)

and that we had not dropped! Blue and white searchlights and all guns were ready to engage. We would have been better close-in because all heavy 40mm flak cannon up and down the harbour and adjacent coasts could not depress their guns sufficiently and reach us. We had only just felt the two mines go when everything happened! A radar directed blue master searchlight came on. There was no searching around the sky. We were in the beam. Our only recourse was violent diving and weaving, and hopefully be elsewhere when the shells started arriving. The Germans are indomitable fighters. Nothing seemed to shake them. White searchlights everywhere tracked the blue. Guns flashed like striking matches. The next instant, shells were bursting everywhere. All hell around and bright red light from flashing shells. Shattering impacts close to the aircraft, with debris titter-tattering all over, through the aircraft skin. A piece of shrapnel entered the tail turret above Jock McCombie's head and lodged in his starboard ammunition pan. At 200 ft we could not "evade" but simply bank out to sea and out of range. Checking for damage, there were holes and rips everywhere, and three large areas of geodetic, perhaps a metre across, were completely wide open and the IFF [identification friend or foe] blown away. We got back but U-Uncle was

in the hangar for repairs for three months.'
Charles Booth, 38 Squadron Wellington radio operator, 1942.

MOONLIGHT BECOMES YOU

The Bomber
White moon setting and red sun rising,
White as a searchlight and red as a flame,
Through the dawn wind her hard way making,
Rhythmless, riddled, the bomber came.
Men who had thought their last flight over,
All hoping gone, came limping back,
Marvelling, looked on bomb-scarred Dover,
Buttercup fields and white Down track.
Cottage and ploughland, green lanes weaving,
Working-folk stopping to stare overhead —
Lovely, most lovely, past all believing
To eyes of men new-raised from the dead.

Anon

'It was the moon, the German nightfighter's moon, that we all thought about most. The moon it was, in all its phases, that governed the tide and tempo of our flying. Sunset had become the evening symbol of the beginning of one's own personal share in the war. The dawn had become the best part of the day. Triumphant over the fenland mists that could shroud a flarepath while your back was turned on the downwind leg of the circuit, the dawn was the thing to warm your bones. You associated it with the incessant roar of engines spluttering and backfiring into a silence so peaceful and positive that it left you drowsy with the relaxation of strain and concentration; so drowsy that it took actual physical effort to slip off your sweaty helmet, slide back the altitude-chilled slide panel in the cockpit canopy and lean out to feel the fresh air on your face and welcome the pleasurable sounds and sights of earth . . .'
Piece of Cake *by Geoff Taylor, RAAF Lancaster pilot.*

I'll press the tit sir, I'll press the tit, sir,
I'll press the tit at the first flak we see.
'Cos I don't like the flak sir, I don't like the flak, sir,
I want nothing but plenty of height for me.

'It was not very pleasant, when you awoke in the morning, to see them gathering up the personal effects of those who failed to return from last night's raid. The normal crew of a Hallybag being seven, three aircraft missing meant 21 wouldn't be around any more . . .' (via Tom Cushing)

(via Tom Cushing)

'A very mixed crew and all nervous as hell.' (via Tom Cushing)

'*Essen* again. You begin wondering how much more it can take. Our crew consists of two Englishmen, a Scotsman, an Irishman, a Welshman, and myself as the Canadian. A very mixed crew and all nervous as hell. It must be remembered that each bomber was really a flying 25-ton bomb just looking for an excuse to blow up. The five tons or so of high-explosives and magnesium flares, plus another three or four tons of high-octane fuel, provided the ideal mixture for a violent explosion when hit in the right place by an explosive bullet or shell. We were losing too many of our friends. It was not very pleasant, when you awoke in the morning, to see them gathering up the personal effects of those who failed to return from last night's raid. The normal crew of a Hallybag being seven, three aircraft missing meant 21 wouldn't be around any more. New replacements would soon arrive and fill these empty beds. And so the war goes on!'

Pilot Officer J. Ralph Wood, DFC, CD, RCAF, 76 Squadron Halifax navigator, Essen, 1942.

MILLENIUM

'The force of which you form a part tonight is at least twice the size and has more than four times the carrying capacity of the largest air force ever before concentrated on one objective. You have an opportunity, therefore, to strike a blow at the enemy which will resound, not only throughout Germany, but throughout the world.

'In your hands lie the means of destroying a major part of the resources by which the enemy's war effort is maintained. It depends, however, upon each individual crew whether full concentration is achieved.

'Press home your attack to your precise objective with the utmost determination and resolution in the foreknowledge that, if you individually succeed, the most shattering and devastating blow will have been delivered against the very vitals of the enemy. Let him have it – right on the chin.'

Air Marshal Arthur Harris, C-in-C RAF Bomber Command on the eve of the first 1,000-bomber raid, 30 May 1942.

Harris was of the opinion that saturation bombing would defeat Germany. On 28/29 March 1942 234 bombers, mostly carrying incendiaries, had bombed Lübeck. Eight bombers were lost, but 191 aircraft claimed to have hit the target. About half the city, some 200 acres, was obliterated. For four consecutive nights, beginning on 23/24 April, Rostock was devastated by incendiary bombs, and by the end only 40 per cent of the city was left standing. The outcome led to Harris wanting to send more bombers to German cities, and a force of 1,046 aircraft was sent to Cologne for the 30 May Millenium raid. Of these, 599 were Wellingtons, and 338 Stirlings, Halifaxes, Manchesters, and Lancasters. No fewer than 367 of the aircraft came from Operational Training Units (OTUs). For 98 minutes a procession of bombers passed over Cologne. Some 898 crews claimed to have reached and attacked the target. They dropped 1,455 tons of bombs, two-thirds of them incendiaries. More than 600 acres of the city were destroyed. In all, 40 bombers and two Intruders were lost, a 3.8 per cent loss rate; 116 were damaged, 12 so badly that they were written off. The fires burned for days, and 59,100 people were made homeless.

'. . . a force of 1,046 aircraft was sent to Cologne for the 30 May Millenium raid. 599 of the aircraft were Wellingtons, and 338 Stirlings, Halifaxes, Manchesters and Lancasters.' (Bart Ryjnhout and BAe)

'Bomber Command had gathered up every bomber that could fly, even the Whitleys, some of which had been retired to submarine patrol over the bays in southern England.' (Bill Cameron via Steve Adams)

'We took part in the first three 1,000-bomber raids, the first being over Cologne on 30 May 1942. This was my 17th op and we really pranged the target, leaving it looking like red-hot embers of a huge bonfire. Within an hour and a half 1,445 tons of bombs were unloaded – two-thirds of them incendiaries – and 600 acres of the city were devastated. With all these aircraft over the target during a short period of time, one wonders how many may have collided! Bomber Command had gathered up every bomber that could fly, even the Whitleys, some of which had been retired to submarine patrol over the bays in southern England. As a matter of fact the Jerry submarines soon learned that it was wiser for them to stay on the surface and shoot it out with the slow-moving Whitleys than to submerge and risk damage from the aircraft's depth charges.

'Flight Sergeant Tackley and crew took part in the second 1,000-bomber raid, over Essen on 1 June 1942. This was another spectacular raid, reminding one of paintings of the Great Fire of London. It was a vision of hell, a vision I would never forget. There was plenty of opposition over this target, which is part of the industrial centre of Germany. Those long cold hours sweating it out in the navigator's and bomb-aimer's compartment always chilled you through, even with heavy flying boots, extra socks, thick gloves, and your flying suit. The usual "cold sweat" didn't help much, either.'

Pilot Officer J. Ralph Wood, DFC, CD, RCAF, Halifax navigator, 76 Squadron, RAF Middleton St George, Durham. Some 956 aircraft were despatched to Essen, including 347 from the OTUs. 31 aircraft (3.2 per cent) failed to return.

FIGHTING FIRE WITH FIRE

'The Germans entered this war under the rather childish delusion that they were going to bomb everybody else and nobody was going to bomb them. At Rotterdam, London, Warsaw, and half a hundred other places, they put that rather naive theory into operation. They sowed the wind, and now they are going to reap the whirlwind.

'We cannot send a thousand bombers a time over Germany every time as yet, but the time will come when we can do so. Let the Nazis take good note of the western horizon, where they will see a cloud as yet no bigger than a man's hand, but behind that cloud lies the whole massive power of the United States of America.

When this storm bursts over Germany they will look back to the days of Lübeck and Rostock and Cologne as a man caught in the blasts of a hurricane will look back to the gentle zephyrs of last summer. It may take a year, it may take two, but to the Nazis the writing is on the wall. Let them look out for themselves. The cure is in their own hands.

'There are a lot of people who say that bombing can never win a war. Well, my answer to that it has never been tried yet and we shall see.

'Germany, clinging more and more desperately to her widespread conquests, whilst even seeking foolishly for more, will make a most interesting initial experiment. Japan will provide the confirmation.'
Air Marshal Arthur Harris.

'As soon as we beat England, we shall make an end of you Englishmen once and for all. Able-bodied men and women will be exported as slaves to the Continent. The old and weak will be exterminated. All men remaining in Britain as slaves will be sterilised; a million or two of the young women of the Nordic type will be segregated in a number of stud farms where, with the assistance of picked German sires, during a period of 10 or 12 years, they will produce annually a series of Nordic infants to be brought up in every way as Germans. These infants will form the future population of Britain . . . Thus, in a generation or two, the British will disappear.'
Walter Darre, German Minister of Agriculture, April 1942.

'On 5 June I completed my 20th op, to *Essen*. It was a hot one and they were ready for us. The damn flak was like lightning flashing in daylight all about us as the searchlights grabbed us over the target. The shell bursts made a squeaky, gritty noise. The smell of cordite was strong, and you had the feeling that someone was underneath kicking your undercarriage, keeping time with the bursts. We were glad to get back without too much damage. A night on the town would sure look good, even though we sometimes missed the last bus back to our base. But then there was always the air raid shelter or a convenient haystack as sleeping accommodation. We were young, and the first early morning transportation back to the squadron would suffice.'
Pilot Officer J. Ralph Wood, DFC, CD, RCAF, Halifax navigator, 76 Squadron, RAF Middleton St George, Durham.

'That was bad. We lost 52 on that raid. Went in at 12,000 ft, got hit and damn near fell to pieces. Went down to 2,000 ft and sort of stumbled home at about 90 mph. Don't really know how we got home.' (Dr Colin Dring)

'That was bad. We lost 52 on that raid. Went in at 12,000 ft, got hit and damn near fell to pieces. Went down to 2,000 ft and sort of stumbled home at about 90 mph. Don't really know how we got home. All my crew were English. We used to have some pretty wild arguments about the States staying out of the war. After that night over Bremen we argued but we never really got mad any more. Going through something like that brings you pretty close.'
Sergeant Harris B. Goldberg, born in Boston, USA, who trained as an air gunner in the RCAF and in October 1941 had arrived in Scotland. He and his crew of a Wellington flew the 1/2 June Millenium raid on Essen and the 25/26 June raid on Bremen,

which was attacked by 1,006 aircraft, 102 of which were Wellingtons borrowed from Coastal Command, and 272 were from the OTUs. Forty-nine aircraft (5 per cent) failed to return. (After flying 273 operational hours in the RAF, and surviving a crash in a 'Wimpy' in the Sinai desert in November 1942, he transferred to the 8th Air Force).

'We were briefed to fly operations to Munich on 24 April along with 15 other crews. The bomb bay was filled with incendiaries, 136 x 36 lb, and 6 x 500 lb "J"-type clusters. Taking off at 20:50 hrs, we headed south, crossing the Sussex coast near Selsey Bill. The Dutch coast was identified on the radar (H_2S), then we headed deep into southern Germany before turning on a north-easterly direction towards Munich. There was a long wait as the target was identified and the markers, bright coloured flares, were dropped. Thoise carrying out this work were Wg Cdr Leonard Cheshire, Sqn Ldr Dave Shannon, and Flt Lt R.S. Kearns, all from 617 Squadron and flying the Mosquito. Looking down at them from our higher altitude, I wondered at the time who on earth I was watching fly so close to the ground, as just about every gun available to the defence force was firing at them . . . Later in the year Wg Cdr Leonard Cheshire (then Gp Capt) was awarded the VC, and the Munich raid featured largely in the citation.'

Sergeant Roland A. Hammersley, DFM, 57 Squadron Lancaster WOP-AG.

"I'm afraid we shall have to leave building the new wing until after the war."

(Punch)

The Master Bomber

He led that crucial raid without which our war was lost
Or, even if won, then at far more fearful a cost.
His flying days were o'er when he and I first met,
I calling at the old pre-fab: where, with typewriter set,
He planned, unassuming man: no flash "war-hero" looks,
Retired Air Commodore, yet nought of rank would show
As on equal terms we chatted, I but an ex-NCO.
Once he complimented me on some verse I had composed
And confided in me regarding a book that he proposed.
Several times I was to call, natter a while, and leave him at his door,
But then he passed, precluding my calling any more.
I have wondered. In his hours of writing and remembering alone
Did he once more hear that four-fold "Merlin" drone?
Recall intercom talk? "Some poor sod's bought it out to port!"
"That bloody gauge's still reading nought!" "Cup of coffee, sport?"
Then: "Half an hour to run! Turn on to O-four-O!"
And, did he in retrospect look down on Peenemünde's awful glow,
Seeing each bomb-load pulverising his target there below?
(Five and a half hundred 'planes on this, his greatest show)
And, as he wrote of it, did ghosts of they with whom he flew
Whisper odd, half-forgotten details which only they knew?
And did old Bomber-boys, having pre-knowledge of his impending fate
Gather and escort the Master Bomber up to Valhalla's gate?

An appreciation of Air Commodore John Searby, DSO, DFC, by G.R.J. Miles.

'If an enemy bomber reaches the Ruhr, you can call me Meier!'
Hermann Goering.

The Battle of the Ruhr was fought between the nights of 5/6 March and 28/29 June 1943. Some 26 major raids were launched on targets in or near the Ruhr, including three on Berlin. By the end of June 34,705 tons of bombs had been dropped by Bomber Command on the Ruhr at the cost of 628 aircraft.

'If I could hermetically seal off the Ruhr, if there were no such things as letters or telephones, then I would not have allowed a word to be published about the air offensive. Not a word!'
Hermann Goebbels, Nazi Propaganda Minister, diary entry.

CHASTISE

'Nearly all the flak had now stopped, and the other boys came down from the hills to have a closer look to see what had been done. There was no doubt about it at all – the Möhne Dam had been breached and the gunners on top of the dam, except for one man, had all run for their lives towards the safety of solid ground; this remaining gunner was a brave man, but one of the boys quickly extinguished his flak with a burst of well-aimed tracer. Now it was all quiet,

'There was no doubt about it at all – the Möhne Dam had been breached and the gunners on top of the dam, except for one man, had all run for their lives.' The Battle of Britain Memorial Flight Lancaster re-enacts the raid on its 50th anniversary, at Derwent in the Lake District, May 1993. The Derwent Dam was used by 617 Squadron crews to practise on before the actual raid. (Author)

except for the roar of the water which steamed and hissed its way from its 150 ft head. Then we began to shout and scream and act like madmen over the R/T, for this was a tremendous sight, a sight which probably no man will ever see again.

'Quickly I told Hutch to tap out the message, "Nigger", to my station, and when this was handed to the Air officer Commanding there was (I heard afterwards) great excitement in the operations room. The scientist jumped up and danced around the room.

'Then I looked again at the dam and at the water, while all around me the boys were doing the same. It was the most amazing sight. The whole valley was beginning to fill with fog from the steam of the gushing water, and down in the foggy valley we saw cars speeding along the roads in front of this great wave of water, which was chasing them and going faster than they could ever hope to go. I saw their headlights burning and I saw water overtake them, wave by wave, and then the colour of the headlights underneath the water changing from light blue to green, from green to dark purple, until there was no longer anything except the water bouncing down in great waves. The floods raced on, carrying with them as they went – viaducts, railways, bridges and everything that stood in their path. Three miles beyond the dam the remains of Hoppy's aircraft were still burning gently, a dull red glow on the ground. Hoppy had been avenged.'

Enemy Coast Ahead, by Wg Cdr Guy Gibson, VC, DSO, DFC, 617 Squadron, who was awarded the Victoria Cross for his leadership on the momentous Ruhr dams raid, 16/17 May 1943. Results: Möhne and Eder dams breached. Sorpe badly damaged. Eight Lancasters were lost, 53 men were killed and three were captured.

RUHR DAMS RAID 'OPERATION CHASTISE' 16/17 MAY 1943

Fifty years on; Flt Lt Joe McCarthy, USA, who flew ED923 'T' to the Sorpe; and Flt Lt Les Munro, RNZAF, who flew ED921 'W' but whose aircraft was damaged by flak over Vlieland on the Dutch coast on the way in and was forced to abort. (Author)

617 Squadron, RAF Scampton, Lincolnshire. 5 Group.
Primary targets: Möhne, Eder and Sorpe dams. Secondary targets; Schwelm, Ennerpe and Diemi dams.

Aircraft (Lancaster B.III)	*Captain*	*Target*	*Remarks*
ED864 B-Beer	Flt Lt David Astell DFC	Möhne	Shot down en route to the dam
ED886 O-Orange	Flt Sgt W.C. Townsend	Ennerpe	
ED906 J-Johnny	Flt Lt David J.H. Maltby DFC	Möhne/Eder	
ED910 C-Charlie	Plt Off Ottley	Lister	Lost en route to the dam
ED918 F-Freddy	Flt Sgt Ken Brown, RCAF	Sorpe	
ED923 T-Tommy	Flt Lt Joe McCarthy, USA	Sorpe	
ED925 M-Mother	Flt Lt J.V. 'Hoppy' Hopgood	Möhne	Lost at the Möhne Dam
ED929 L-Love	Flt Lt Dave J. Shannon DFC, RAAF	Möhne/Eder	
ED934 K-King	Flt Sgt Byers, RAAF	Sorpe	Lost
ED937 Z-Zebra	Sqn Ldr Henry E. Maudslay	Möhne	Lost: blown up by the blast of his own bomb
ED865 S-Sugar	Plt Off Burpee, RCAF	Sorpe	Lost: killed by flak
ED887 A-Apple	Sqn Ldr Melvyn 'Dinghy' Young DFC	Möhne/Eder	Shot down over the Dutch coast on return
ED909 P-Popsie	Flt Lt Mick Martin DFC, RAAF	Möhne	Returned after bombing Möhne
ED912 N-Nuts	Flt Lt L.E.S. Knight, RAAF	Möhne/Eder	(KIA 15.9.43)
ED921 W-William	Flt Lt Les Munro, RNZAF	Sorpe	Damaged by flak over Vlieland on Dutch coast on the way in and forced to abort
ED924 Y-Yorker	Flt Sgt Anderson	Sorpe	
ED927 E-Edward	Flt Lt Barlow, RAAF	Sorpe	Lost
ED932 G-George	Wg Cdr Guy Gibson DSO DFC	Möhne/Eder	(KIA 19/20.9.44 in a Mosquito). His crew on the Dams raid were all lost on 15.9.43)
ED936 H	Plt Off Allan Rice	Sorpe	Hit the sea before crossing enemy coast and aborted, flying home on two engines. (KIA 20.12.43)

(Author)

GOMORRAH

'In spite of all that happened at Hamburg, bombing proved a comparatively humane method.'
Air Chief Marshal Sir Arthur Harris.

On 24/25 July 1943 791 bombers dropped over 2,200 tons of bombs on Hamburg. 'Window' was used for the first time, and rendered the German radar system ineffective. Firestorms devastated the city and fires were still burning 24 hours after the raid. On the nights of 27/28 and 29/30 July and 2/3 August, mass raids were again made on the city. In total, some 3,095 sorties were made and almost 9,000 tons of bombs were dropped. Over 42,000 people are thought to have been killed.

'A wave of terror radiated from the suffering city and spread throughout Germany. Appalling details of the great fires were recounted, and their glow could be seen for days from a distance of 120 miles. A stream of haggard, terrified refugees flowed into the neighbouring provinces. In every large town people said: "What happened to Hamburg yesterday can happen to us tomorrow". Berlin was evacuated with signs of panic. In spite of the strictest reticence in the official communiques, the Terror of Hamburg spread rapidly to the remotest villages of the Reich.

'Psychologically the war at that moment had perhaps reached its most critical point. Stalingrad had been worse, but Hamburg was not hundreds of miles away on the Volga, but on the Elbe, right in the heart of Germany. After Hamburg in the wide circle of the political and military command could be heard the words: "The war is lost".'
Adolph Galland.

It was not, of course, and night after night for almost two more years the Grim Reaper continued to strike. The worst night of all was on 30/31 March 1944, when 795 bombers were despatched to Nuremburg:

BLACK FRIDAY

'. . . Instead of the bomber stream being 5 miles wide it was more like 50. Some had already been shot down and before I reached the far side of the stream they were being shot down on my left. Masses of Window were being tossed out of the bombers, which also jammed our radar. We tried three times, but each time came up below a bomber, the rear gunner spotting us the third time, his tracer coming uncomfortably close whilst his pilot did a corkscrew. It was hopeless, we were doing more harm than good. Ahead bombers were being shot down one after another, some going all the way down in flames, some blowing up in the air, the rest blowing up as they hit the ground. I counted 44 shot down on this leg to Nuremburg. What was happening behind I could only guess.

'. . . I was inwardly raging at the incompetence of the top brass at Bomber Command.'
Flt Lt R.G. 'Tim' Woodman, Mosquito pilot, 169 Squadron, 100 Group Bomber Support.

'During my leave in Canada, Bomber Command suffered its worst defeat of the war. This was the Nuremburg raid on the night of 30/31 March 1944, with the loss of more aircrew than were lost in the entire Battle of Britain. Over 800 experienced aircrew were dead, wounded or missing or prisoners of war. Bomber Command's maximum acceptable rate of loss was considered to be 200 four-engined bombers with their seven-man crews a month. Not much to look forward to on my return.'
Pilot Officer J. Ralph Wood, DFC, CD, RCAF navigator.

Ninety-four bombers were lost, another 12 crashed on return, 59 were badly damaged, and 745 aircrew were missing. This was a missing rate of 11.8 per cent and a total casualty rate of 20.8 per cent.

ODE OF COMMODE

One hundred mighty bombers, returning from a
raid,
Their target, the most distant that they had ever
made;
Diverted to an airfield in the region of Exmoor
Without a place to sleep them save on a hangar floor

And there, with palliasses as apology for beds,
Seven hundred flyers lay down their weary heads.
They had sucked in oxygen for many, many an hour
Which now released as tummy-gas was noxious in odour
And as they'd also eaten hasty, gaseous food
The atmosphere was soon described as anything but good;
Thus tired, exhausted flyers combined to make a rattle,
Said to confuse some passer-by in thinking it was battle.
'Twas just a wartime incident which never made the press
And I would never know about but for a friend's distress
One who was a flyer in the story that I tell,
Who, having damage to his back, experienced some Hell;
Of course the doctor drugged him to modify the pain,
This drug has a side-effect; hallucinates the brain.
Now, in this semi-conscious state, his mind went to-and-fro
To where he was reliving that night so long ago
When sleeping in discomfort with a hangar for abode;
This time, they had one toilet; his invalid commode!
And seven hundred backsides all needing to discharge
Within this bedside "office" which was not very large.
He viewed the scene with humour for, as one might expect,
Each and every backside desired that it be next;
Seven-hundred bottoms all fighting to let free
Provided him a vision, it seems he watched with glee.
I understand it ended in an utter mad confusion,
He 'woke in helpless laughter and gone was his illusion
But he can still remember the details of that sight
With seven hundred flyers in toilet seeking fight
He told me the story, knowing I'm the rhyming kind
And, like most other poets, have a naughty mind
So I've employed my talent to put his tale to verse,
Confirming all suspicion that I can be perverse.
Now I have dreamed of wondrous sights whilst sleeping like a log,
But ne'er a horde of backsides in combat for one bog.

Jasper Miles

LITURGY FOR THE LIBERATOR

A Lib Leaving Malta For Gib

They say there's a Lib leaving Malta for Gib,
Heavier than ever before,
Tight to the turrets with terrified troops,
Fifty or sixty more.
There's many a Hun with a gun in the sun

'. . . At first light the following day eight Liberators from 59 Squadron arrived to protect the convoy. Our total flying time on this mission was 15 hrs 35 min.' (Author's collection)

As they trundle back home to the Rock.
In case of brake failures they're glad of the Sailors
As they ditch at the end of the dock!
Airmen's Song Book

Consolidated B-24 Liberators were used extensively by RAF Coastal, Special Duties, Transport and Bomber Commands in the Mediterranean, and Middle and Far East.

'We obviously hit a munitions supply train, as the explosions on the ground were like a firework display and even at 15,000 ft we had the feeling of flying right through an inferno. Most raids were flown at around 12,000 to 15,000 ft. There was no point in going higher and losing accuracy in bombing. Over the target we had to stagger our bombing heights to lessen the chance of collisions and sometimes we had to descend to, say, 8,000 ft if we had been allocated one of the lower levels. This meant that instead of being able to get the hell away from the target, flak and the prowling night fighters, we had to climb at slower speed to get back to bomber-stream height.

'Also on raids to Austrian targets we had to get back over the Alps . . . After bombing a rail junction and marshalling yard near Innsbruck, we had to climb immediately in order to clear the mountains. Had we been only slightly off course we could not have made it, due to the higher peaks on either side. This night we entered cloud as we turned from the target for home. The cloud was coloured red from the ground fires and there was no visual contact with the mountain peaks all around. As we climbed, the Liberator seemed slower than usual and we seemed to be heading directly for a mountain. I turned the electronic boost control into "Emergency position 10" and the Lib seemed just to jump upwards. At that moment we broke cloud into a magnificent scene of the Tyrolean Alps in bright moonlight. The panorama below was breathtakingly beautiful, and a strong contrast to the scene of death and destruction we had just created only a few miles away.'

Deryck Fereday, pilot, 178 Squadron, night raids in a Liberator from Italy, March 1945.

'When we arrived, one ship was on fire and ships were being attacked by a Focke-Wulf 200 using high-level bombing technique . . . We got on the tail of a Focke-Wulf which was going to attack, but the gunners on the ships let all hell loose. They fired at the Focke-Wulf 200 and ourselves but caused the enemy aircraft to swerve just as he was dropping his bombs. He missed his target altogether, and to our amazement all seven Condors broke off the action and scurried for home . . . At first light the following day eight Liberators from 59 Squadron arrived to protect the convoy. Our total flying time on this mission was 15 hrs 35 min.'

John Branagan, B-24 WOP/AG, 59 Squadron RAF Coastal Command, Aldergrove, Northern Ireland, convoy protection duty, 29 July 1943.

'On 5 April 1943, when escorting convoy HX231, we sighted and attacked our first U-boat. We dropped a stick of four depth charges close to the U-boat, and the result was a probable kill. Five days later we attacked two more U-Boats. This operation lasted 15 hours. As the U-boats stayed on the surface and fired a 3.7mm gun from this position in the stern of the conning tower, there was always the strong possibility of being shot down. In 1943 an 0.5 in machine-gun was mounted in the Perspex nose compartment of the Liberator to upset the U-boat gun crews. We were lucky, but at least two Liberators were believed to have been brought down while making such attacks.'

Edward Bailey, WOP/AG 120 Squadron, RAF Coastal Command, North Atlantic, spring 1943.

'The 2nd pilot handed me a message in code which read: "I am attacking enemy submarine on surface". My Morse was not exactly perfect because of the noise of the guns and the swerving of the aircraft. Once more we circled the U-boat and dropped our second salvo of charges. The German crew scrambled out through the conning tower and abandoned ship. Seconds later it exploded and broke in half, leaving about 30 survivors in the sea.'

John Branagan, Liberator WOP/AG, 59 Squadron RAF Coastal Command, south of Greenland, 17 October 1943.

Coastal Command Liberators, Wellingtons, Hudsons, Catalinas, and Sunderlands waged an unremitting and initially unrewarding war against the German U-Boat menace. Through technological developments such as the Leigh light and radar, but mostly through the determination and courage of its aircrews, Coastal Command played a major role alongside surface forces in defeating the submarine threat.

By May 1945 the Command consisted of seven groups with a total of 24 squadrons of long-range anti-submarine-warfare (ASW) aircraft, flying boats and strike aircraft, which together had accounted for 191 U-boats over the previous six years.

Other RAF bomber squadrons, some of them equipped with Liberators, also made attacks on German submarines.

'Towards the end of our patrol we spotted a U-boat on the surface, travelling at about eight knots. It immediately opened fire with everything it had . . . the bomb bay doors were open when I saw tracer shells passing under the wings. We dived down to attack . . . We went in just above the altitude for accurate flak at about 24,000 ft. This was too high for such a small target, so we were instructed to drop one bomb as a marker, make the necessary corrections on the bombsight and go around again and drop the rest of the bomb load. On approaching the target we were immediately subjected to deadly accurate flak, and after a brief discussion with our bomb-aimer we decided to use the marker bombs dropped by two Libs who were ahead of us. The bomb run was very "dicey". When the bomb doors were closed, full power was applied and a steep climbing turn executed. My hands were shaking so violently that I couldn't make my log entry for several minutes.'

Denis Allen, Liberator flight engineer, 40 Squadron, describing a daylight raid on a wharf at Arsa, 1945.

In the Far East the fall of Malaya and Singapore had been a major blow, and the RAF was hard-pressed to offer any effective resistance to the well equipped and tenacious Japanese invasion forces. After reorganizing and re-equipping with more modern aircraft, including Spitfires, Hurricanes, Beaufighters, Thunderbolts, and Liberators, the RAF fought back, often in the most gruelling conditions imaginable, from its bases in India, Ceylon and Burma, and in conjunction with the USAAF turned the tide of war against the Japanese. By 1945 the RAF element of the Air Command SE Asia consisted of ten groups with 75 squadrons.

'Morale was low at the time. There was no mail from home and we could not get any spares for

'Towards the end of our patrol we spotted a U-boat on the surface, travelling at about eight knots. It immediately opened fire with everything it had . . . the bomb bay doors were open when I saw tracer shells passing under the wings. We dived down to attack . . .' (Author's collection)

the aircraft. Even so, we managed to get about four out of six Liberators ready for operations to Ramree and Akyab Island . . . We received Mark VI Liberators and later Mark VIIs and started a gradual build-up of spares and a regular supply of beer. At times the squadron flew with American Liberators on operations. They could never understand how we managed to take off with such heavy loads of fuel and bombs.'
Flight Sergeant Stanley Burgess, fitter, 159 Squadron, Salbani, India, 1942.

'It was a well planned operation with a diversionary raid laid on at high altitude to soak up any fighter opposition. We six flew at low level over the Golden Pagoda, which was our initial point for the mining operation. At timed intervals each Liberator dropped a string of mines almost on the Pagoda's doorstep, all the way along the river and almost up to Elephant Point at the river's mouth. We flew position No. 5 in the dropping order. All six aircraft dropped their mines and headed straight out to sea for home . . . An inferno erupted on what was a peaceful night. Right in the middle of it all a Liberator had been hit. It caught fire, keeled over and went in with one big explosion. It was only on our return, about 100 miles from base, when we broke radio silence and the call signs came through, that I realised I had witnessed the destruction of my crew with whom I had flown 23 operations.'
John Hardeman, Liberator rear gunner, 159 Squadron, describing a minelaying operation in the river approaches to Rangoon on the night of 29 December 1944.

'We six flew at low level over the Golden Pagoda, which was our initial point for the mining operation.' (The late Gp Capt Jones)

DUFF GEN

Press on regardless – never mind the weather
Press on regardless – it's a piece of cake
Press on regardless – we'll all press on together
'Cos you're bound to see the Dummer or the
Steinhuder Lake
(Tune 'Poor Joey')

'The meteorology reports were very important to the success of our operations. The met section, at times, left a lot to be desired. We called the meteorology information, "met gen", which usually turned out to be one of the following: "pukka gen", meaning "good information", or "duff gen", meaning "bad information".'
Pilot Officer J. Ralph Wood, Whitley navigator, 102 Squadron.

'First the port outer went. "What do we do skipper?", the crew said. I said: "Press on regardless". Then the starboard outer went. "What do we do now skipper?", they said. I said:

(Punch)

"And pray, what might 'Nil excretum Taurus' mean."

". . . and this is the undercarriage control."

"Whaddya mean — Back a bit?!"

"Calling C for Charlie. Calling C for Charlie. All right — sulk if you want to."

"Press on regardless". Then the port inner went. "What the **** do we do now skipper?", they said. I said: "Say after me the Lord's Prayer . . !"'
Bomber pilot.

'"Butch Harris is in the boob at Dulagluft," a bunk neighbour tells you. Speechless, you ponder over the prospect of Air Chief Marshal Harris, AOC Bomber Command, being shot down, captured, and interrogated by the Luftwaffe. If there's one man in the world the Germans would most like to have in a cell, short of Churchill, it surely must be Harris.

'"Jesus," Ruminates your equally awe-struck informant, "poor old Butch."

"One of our mob," comments a sardonic Australian voice.

'The reactions to the quickly-discredited rumour are fairly typical of those of all the survivors here of the bombing policy prosecuted so ruthlessly by Harris and immortalized so cynically by his own crews in the imperishable exhortation: "Press on, rewardless!"

'Maybe the Bomber Command crews still flying on operations have their own ideas. As far as most of us here are concerned, there is a kind of savage, blasphemous pride to be felt in having been one of Butch's boys.'
Geoff Taylor RAAF, PoW, Stalag IVb, Mulhburg-on-Oder.

'A quickening of interest, a sort of tensed leaning forward to hear better, among the pilots, who a few minutes earlier had been listening abstractedly to the planning details, greeted the intelligence officer's commencement of the current summary of the location and weight of the flak, searchlight, and nightfighter defences of Germany as it was likely to affect us.

'There was, it seemed, nothing to worry about.

'"Should be a piece of cake, chaps," he said, smiling at the cynical grunts and mirthless snorts of laughter from the old hands.'
Geoff Taylor, RAAF, pilot of Z-Zebra, Piece of Cake, *George Mann, 1956.*

'Flying never killed anyone. Crashes do, though.'
Anon

'The Luftwaffe now rates Mosquitos on night attacks so high that when a Nazi pilot shoots one down he is allowed to count it as two.'
British Newspaper column.

SHOOTING A LINE

I'll shoot 'em down sir, I'll shoot 'em down, sir,
I'll shoot 'em down if they don't shoot at me.
Then we'll go to the Ops Room and shoot a horrid line, sir,
And then we'll all get the DFC.

'There was a story going around about a Whitley crew becoming lost and running out of fuel. The pilot set the automatic pilot and told his crew to bale out. This they did, with the exception of the tail gunner, whose intercom had become disconnected, and he failed to hear the order. A short time later the aircraft made a remarkably good landing on a sloping hill in Scotland. The tail gunner, upon vacating his turret, commented loudly on the pilot's smooth landing. You could almost see him passing out when he discovered he was the only occupant of the aircraft.'
Pilot Officer J. Ralph Wood, DFC, CD, RCAF

'Push the stick forward, the houses get bigger. Pull it back, they get smaller. Pull it too far back and they get bigger.'
Anon.

From the Line Book

'You could almost hear the tracer sizzling as it went past.'
Flying Officer Willie P. Rimer, Mosquito pilot.

'I've flown a Beaufighter at nought feet on instruments.'

'We do our air tests on ops.'

'We've been doing our ops so fast that I've lost count!'

'I'm always glad to go on ops; it takes my mind off the war.'

'I remember one night, I was shooting-up road signs . . .'

Navigator: 'Did you hit the drogue?'
Pilot: 'No, I left it this time. I shot so many away on the gunnery course, they're getting a bit short.'

'We've never been known to come back unserviceable unless the 'plane's practically in two pieces.'

(Punch)

". . . and light but accurate flak."

"I wasn't sure whether it was one of theirs or one of ours, so I only gave it a short burst."

". . . and then with a final twenty-minute burst, I more or less dismantles the entire aircraft."

"We *went down to a few* inches."

'When things get rough I light my lighter, throw it out and get a pinpoint.'

'The only time I get to sleep is when I'm on ops.'

Pilot on his low-flying prowess: 'I used to cut the grass for my wife when we lived out.'

The Cadet on his first solo had engine trouble, and was forced to land his Tiger Moth in a field. He was followed a little later by his instructor overhead, who, seeing his predicament in the tiny field, thought: 'If he can land in that small field, so can I'. The instructor brought the Tiger in but made a hash of it and crashed in a heap of tangled wreckage. Extricating himself with as much aplomb as he could muster, he said to the onrushing cadet: 'How on earth did you get *your* Tiger Moth into this field?' The cadet replied: 'Through that gate over there sir!'
Anon.

'Our job – 681 Spits and 684 Mossies – was to give notice of any Jap reinforcing after Slim's 14th "Forgotten Army" had foiled the Jap invasion of India and was following them across the Yomas into the plains of northern Burma. We did this by photographing every town, every tin-pot port, every 'drome, roads, and every inch of the railway system. We could keep track of every railway wagon the Japs had. If the Japs mislaid one we could have told them where it was. We had a line shoot. The old stations had toilets at each end of the platform, so we said: "The Jap stationmasters can't go to the crapper without our knowing".'
M. Howland, RAAF, 684 Squadron pilot.

Pilot to ground crew: 'I could not drop my bombs over the target'. After the ground crew had listed all the possible reasons except for lateral thinking, the pilot answered: 'I'll tell you all why. We were upside down.'
Anon.

'Reminds me of the time I sank the *Tirpitz*,' comments a Spitfire pilot. 'Just one pass, of course, old boy.'
PoW commenting on a sketch in the Volkischer Beobachter *and* Allgemeine Zeitung *of a Fw 190 at low level, strafing a British battleship.*

BLESS 'EM ALL

There's many a troopship just leaving Bombay,
Bound for old Blighty's shore
Heavily laden with time expired men
Bound for the land they adore
There's many an airman just finished his time,
And many a twerp signing on
They'll get no promotion
This side of the ocean
So cheer up my lad, Bless 'em all.

'The high spot of the RAF activities during the Second World War occurred at RAF Castle Bromwich in 1943. When airmen heard a 'plane

'When the Sunderland had stopped the C-in-C said: "I bet I had you worried sergeant?' Well son, I'm full of tricks."' (Shorts)

landing late at night, they assumed it was one of many Spitfires tested there. Switching on an Aldis lamp, however, Aircraftman R. Morgan observed that it was a German bomber. As it taxied down the runway he expressed the intention of having a crack at it with the Lewis gun, and went off to get permission.

'While the German 'plane revved its engines, Aircraftman Morgan tried to ring through to control. "We had to crank like fury on the field telephone for permission to fire," he said. By the time he got through the 'plane had taken off and was en route for Germany.'

The Least Successful Attempt to Shoot Down Enemy Planes, from The Book of Heroic failures, *by Stephen Pile.*

'The C-in-C, a Lancaster pilot, took the left-hand seat of the Sunderland, and a young sergeant pilot took the right-hand seat. The C-in-C took off from Calshot water and headed inland. He approached the runway of an airfield, dropped the flaps, throttled back, and then, much to the relief of the sergeant (who, conscious of his rank, had remained tight-lipped throughout the descent), was relieved to see the C-in-C open up the throttles and set course for Calshot, whereupon he alighted. When the Sunderland had stopped, the C-in-C said: "I bet I had you worried sergeant? Well, son, I'm full of tricks."

'With that he opened the door and stepped out!'

Anon.

TYPES

Vain type – Undoes five buttons where one would do.

Excitable type – Pants twisted, can't find the hole, tears pants in temper.

Sociable type – Joins friends in piss whether he wants one or not – says it costs nothing.

Timid type – Can't piss if anyone is watching, pretends he has pissed and sneaks back later.

Indifferent type – Urinal being occupied, pisses in sink.

Clever type – Pisses without holding his tool, shows off by adjusting neck-tie at same time.

Frivolous type – Plays the stream up, down and across. Tries to piss on flies as they pass.

Absent-minded type – Opens vest, takes out tie and pisses himself.

Disgruntled type – Stands for a while, grunts, farts, and walks out muttering.

Personality type – Tells jokes while pissing, shakes drop off his tool with a flourish.

Sneaky type – Drops silent fart whilst pissing, sniffs, and looks at bloke next to him.

Sloppy type – Pisses down his trousers into his shoes and walks out with flyhole open and adjusts.

Learned type – Reads books or papers whilst pissing.

Childish type – Looks down at the bottom of the urinal while pissing to watch the bubbles.

Strong type – Bangs tool on side of urinal to knock the drops off.

'For some peculiar reason seemingly inherent in the Lancaster's system I always got too hot, Jock at his navigator's desk a few feet behind me seemed to be perpetually frozen, and Joe, sitting at his radio just forward of the main spar, seemed to roast.

'Don, prowling restlessly back and forth at his flight engineer's panel, rarely seemed to use his jump-seat alongside me and appeared to be oblivious to the temperature.

'Smithy, lying bundled by his bombsight and panels in the cold transparency of the nose, was a Canadian and therefore didn't count, for if he complained about the cold he was promptly told to go back to the frozen north.'

Geoff Taylor, RAAF, Lancaster pilot, author of Piece of Cake *(George Mann, 1956).*

'Our mess parties had a habit of accelerating as the evening advanced. On one occasion an officer rode into the mess on horseback, much to the delight of the inebriated occupants. Not to be outdone, another officer jumped on his motorcycle and rode that into the mess. Then there was the night of the duel. An RAF officer who, unbeknown at the time, was suffering from appendicitis, was throwing beer mugs up in the air, breaking the chandeliers, and at the same time was telling the Canadians present that he didn't want any damn Canadians over here to fight his battles and why didn't we get the hell home. I figured that kind of talk called for a duel, and we

took down the two crossed sabres hanging on the mess wall and started fencing. When I nicked him between the eyes, over his nose, our medical officer, who was black as the ace of spades (fondly referred to as old 23:59 – one minute to midnight), decided it was time to end the duel.'
Pilot Officer J. Ralph Wood, DFC, CD, RCAF, 692 Squadron Mosquito navigator.

'Dining-in nights in sergeants' and officers' messes have always been occasions for high jinks! When the formality of dinner is finished, all manner of games are played. After a dinner in the sergeants' mess at Foulsham mid-1944, and following the usual capers, there was a 250cc DR motorcycle race around the ante-room! Another bit of nonsense was the challenge to hang upside-down by the crook of one's legs from a high open rafter of the mess ante-room and drink a pint of bitter!

'. . . The officers' mess at RAF Marham is an imposing three-storied building built during the 1936 RAF expansion programme. Those cars that were sprinkled around the car park were mainly prewar vehicles such as "E"-type Morrises, early Fords, Hillmans and Standards. One chap had a nice little Austin 7 which was his pride and joy. A dozen of his fellow officers went outside and managed to lift the little car up the steps and into the entrance hall. When the owner blew his top, the culprits carried the little car back down the steps again!

'. . . Having formed up, the parade was brought to attention and "open order" by the adjutant and handed over to the new, nervous, station commander for inspection. His first task was to prepare the parade for an address by the padre. Unfortunately he issued the never-to-be-forgotten command: "Fall out Romans, Catholics, Jews, and other denominations!"'
Happy Memories, *Gp Cdr Jack A.V. Short.*

'What started off as a civilised party developed into a real "thrash". Ossie, the Scots tea planter from Ceylon, climbed on to the bar counter. "Right!", he bellowed, "I think we all need a bit of practice at landing on FIDO [fog investigation and dispersal operation]. I'll be the controller." Newspapers and magazines were confiscated from the lounge, rolled up and placed in almost parallel lines on the highly polished mess bar floor. Tins of Ronseal were sprayed on the papers. "Let's make it realistic," said someone. "Get the feather cushions from the lounge and we'll have 10/10ths feather visibility. The idea was to slide down the polished line between the burning papers and have the air full of feathers at the same time. Ossie bellowed out the time-honoured phrase: "Come in number one, your time is up!" Whereupon the CO whipped down the burning line of newspapers on his bottom and crashed into the stove at the other end. The atmosphere of the burning papers plus the clouds of feathers made the bar almost untenable.

'It could have been worse. They could have burned the place down. At one party a bunch of aircrew from another squadron had found a pile of bricks, sand, and cement. Contractors were building an extension to the mess. The chaps used the lot and bricked up the CO's car, which was parked outside. The contractors had been reported as saying that they hoped they were better flyers than they were bricklayers.'
John Clark, navigator, 192 Squadron Mosquitos, in One Man's War.

'We usually slept until the last minute, then made a mad dash for the mess before the doors closed. Most of us had this timed pretty well, so well that Andy and I decided to upset the pattern by piling as many bicycles as possible on top of the latrine building. Of course, a great many missed their breakfast that morning, including the CO, whose bicycle was also included.'
Pilot Officer J. Ralph Wood, DFC, CD, RCAF, 692 Squadron Mosquito navigator.

'When two 'planes collided and went up in flames, the fire crew ran out of foam and came back to the stores for a further supply. Sergeant Booty, a Norwich man, demanded a voucher. He was lucky to survive! I had a run-in with the Engineering Officer [EO], who rang when I was alone in the stores one day. He wanted to know how many rear tyres we had for our Blenheim aircraft. Now we were out. The aircraft at Attlebridge were running on a concrete runway (it was later tarred, had sawdust rolled into it and then sprayed green camouflage). The rear wheels did not run true, so every time a 'plane landed they became u/s. I told the Engineering Officer that I didn't think we had any. He said I wasn't paid to think. Crash went the phone at the other end. Minutes later the EO arrived, bawled for the key to the rubber store, and off we went. Of course there were no tyres to be seen. I told him I had sent a despatch rider to Horsham St Faith, West Raynham, and Marham to collect a few until we could get a supply from a MU [Maintenance

Unit]. Without a word he went.

'On another occasion I was asked: "Where are the Wellington bombers?" I said we hadn't got any, but the Air Ministry was adamant – they wanted to know if they were still serviceable. Then the penny dropped. They were the plywood dummies in a field along the Norwich road which had been put there as decoys!

'One day we ordered six airscrews from 25 MU for our Bostons. A few days later I got a 'phone call from Dereham asking for transport from the station. Six aircrews, 18 flyers, had turned up! As a result in future all requests for "Airscrews" were changed to "Propellers"!'
Corporal A.E. Orford.

Cheers (A Flyer's Toast)

To the men who turned the spanner, to the men who pulled the wrench,
To the men who did refuelling with the octane's heavy stench,
To the 'genius' with radar, to the man who fixed the gun,
To the services crews in freezing cold when working was not fun,
To those who brought the bombs along and loaded them aboard,
To the artist of the mascot, be it Pluto or a broad,

'To those who brought the bombs along and loaded them aboard / To the artist of the mascot, be it Pluto or a broad.' (RAF Marham and Stanley Burgess)

To the cooks who cooked the dinner, though not always Cordon Bleu,
To the girl who brought the break truck (and what you thought of her?),
To those who spread the bullshit from their office in the warm,
To those who crewed the ambulance in case you came to harm;
For each and every flyer owes a debt he cannot pay
To those who worked upon the ground and sent him on his way.
Jasper Miles

'They were the backbone without whose expertise no operations would have been possible. They worked all hours and in all weathers to ensure that our aircraft were serviceable.' (Corporal Jock Bell, LAC Crown, and Harry Castledine, 226 Squadron.) And on many occasions when we carried out two or three ops in a day they were always on duty to refuel, bomb-up, and rearm the guns and patch up the odd flak damage.' (RAF Swanton Morley and author's collection)

'To those who worked upon the ground and sent him on his way.' Counter Attack being handed over to the RAF by NAAFI canteen assistant Nora Margaret Fish. (RAF Swanton Morley)

GROUND CREWS

For lack of a bolt the engine was lost,
For want of an engine the bomber was lost,
For want of a bomber the crew were lost,
For want of the crew the battle was lost,
For want of the battle our freedom was lost,
And all for the want of a bolt.

'They were the backbone without whose expertise no operations would have been possible. They worked all hours and in all weathers to ensure that our aircraft were serviceable. And on many occasions when we carried out two or three ops in a day they were always on duty to refuel, bomb-up, and rearm the guns and patch up the odd flak damage.

'Very rarely did they grumble, and on the occasions when they did we fully sympathised with them. Their 'Chiefy' was an older man with years of service, and encouraged his lads in every possible way. On the occasional "stand down" we would take them out to one of the local pubs which they enjoyed, and we most certainly appreciated their interest in us and our aircraft, which they considered to be their aircraft. Although our closest association was with our ground crews, we also remember with gratitude the WAAF drivers who took us to our aircraft. We remember the parachute packers, the cooks who provided excellent meals at some very odd hours and, of course, many other trades too numerous to mention. Our successes were due to team effort and esprit de corps.'
John Bateman, Boston gunner.

Now pilots are highly trained people
and wings are not easily won . . .
But without the work of the maintenance man
our pilots would march with a gun . . .
So when you see the mighty aircraft
as they mark their way through the air
The grease stained man with the wrench in his hand
is the man who put them there.

Anon

BRYLCREEM BOYS

'16 August. I have moved over to Wyton so I thought I had better give you the gen. This is a peacetime mess and absolutely wizard. There are two of us to a room about the size of a large office, and as my pilot is not coming over until tomorrow or Friday I have it to myself at the moment. It has two chests of drawers, a built-in wardrobe, two chairs, tables, and reading lamps. The food is terrific; tonight for dinner we had soup, a choice of rabbit or cold meat and salad, plum pie, stewed plums, trifle or rice and plums and cheese and biscuits. We are waited on at table and the messing is only a shilling a day.'
Derek Smith, Mosquito navigator, in a letter home, 1944.

'The RAF used to have eggs and bacon for breakfast. All we got were beans on the second sitting. We had this corporal in our Company who could imitate the Air Raid Warning whistle. Sometimes he would stand in the main doors of the mess and whistle. All the RAF people would come running out and dash into the air raid shelters. Then we would go in and eat their breakfast for them. As we left the mess he would give the "All Clear"!'
David Woodhouse, 70th Suffolk Yeomanry.

'Some nights at our debriefing, Andy Lockhart, my pilot, and I would use up the rum ration of those who preferred their coffee without this additional fortification. This on top of our own ration had the effect of making everything seem great. The only difficulty was bicycling from the ops room to our barracks without going in the ditch. The only time we had bacon and eggs was for a flying breakfast when we went on ops. I used to think they used this as bait to get us to fly. Once in a while we would bicycle around the countryside, looking for farmers to sell us some eggs. We would then have the cook at our mess prepare them for a snack at night after visiting the pub. The eggs kept getting a little tougher, and eventually we realized that the farmers were passing off duck eggs on us.'
Pilot Officer J. Ralph Wood, DFC, CD, RCAF, 692 Squadron Mosquito navigator.

(Punch)

"Right — right — left — steady — steady . . ."

THE SAME AS YESTERDAY

'At the station, with a gut load of sergeants' mess garbage stuck in my crop, I'd hop on my bicycle and pedal into Abingdon for a few beers to wash it down. The mess food was usually mutton and more mutton, kidneys, curried rice and, of

(Punch)

"All right, Farthingale, we give up. Where HAVE you hidden the dinghy?"

course, brussel sprouts. The latter were a kind of miniature cabbage, eaten boiled. At our mess the cook must have boiled them and reboiled them until they emerged a sickly green gob. My RAF friend, who often ate with me, used to mutter "13–8, 13–8, 13–8" whenever we had these brussel sprouts as a vegetable. My curiosity got the better of me, and I asked him what he meant by this "13–8" expression. He said: "Ralph, when I'm served this mess I just have to express my displeasure, but it's not swearing; it's taken from the Bible. Hebrews, 13th Chapter, 8th verse: "Jesus Christ, the same as yesterday, and today, and today, and forever."'
Pilot Officer J. Ralph Wood, DFC, CD, RCAF navigator.

We are the Air-Sea Rescue, no effing use are we;
The only times you'll find us are breakfast, dinner and tea.
And when we sight a dinghy, we cry with all our might:
'Per Ardua ad Astra – up you, Jack, we're all right.'

'A party of ATC cadets was suddenly let loose in the hangar. The novelty of visitors soon wore off as we tried to get on with our work. I was working in the dinghy stowage area and came up for air to see a cadet sitting on the pilot's seat. A rigger had already told the cadet not to mess about with switches or controls. As I rested for a minute or two I heard the cadet ask "What's this?", as he held up the chromium-plated item with flexible tube. He received the perfectly serious reply: "That's to plug the oxygen into". Whereupon he took several deep sniffs and said: "I didn't know oxygen smelt like that".'
LAC W.G. 'Bill' Cooper, electrician II.

'One of my compatriots from Moncton, a navigator, missed Great Britain altogether when returning from a raid. He landed in southern Ireland, which was neutral, and remained there in internment for the rest of the war, enjoying good food and drink while his pay and promotion continued. I still haven't decided whether he was a stupid navigator or a smart operator.'
Pilot Officer J. Ralph Wood, DFC, CD, RCAF, Whitley navigator.

'Making a good landfall on the English coast on our homeward journey always boosted the navigator's morale. By a good landfall I mean approaching the coast and hitting it just about where you were supposed to – right on track. I recall my pilot asking occasionally: "Where are we?" I'd shove a map or a chart in front of him, pointing wildly to any spot over the North Sea or Germany, depending on the occasion. This having satisfied him, I returned to my plotting table to work out our actual position undisturbed.'
Pilot Officer J. Ralph Wood, DFC, CD, RCAF, Whitley navigator.

'Bedale, the squadron village, was a couple of miles distant and was a great place for relaxation. It boasted eight pubs, which, during off-duty period, could mean a drink at each. I think the favourite was the the Green Dragon, which had a cross-eyed barmaid named Kathy who played darts for her knickers or your underpants! One or two, but very few airmen had Kathy's knickers displayed on the bedroom wall. I think she had acquired a job lot, as they were all the same, emblazoned with traffic-light motifs in the appropriate places.'
Basil Craske, Whitley pilot, 10 Squadron.

'We shared quarters with a Free French crew which included Sous Lt Roger Masson, an air

(Punch)

". . . Won't keep you a minute. The Skipper's sea-sick."

gunner, whose naviagtor was Capitaine Pierre Mendes-France, who for a short while after the war became Prime Minister of France. Mendes-France habitually read until the small hours of the morning, but one night Roger, whose bed was opposite, politely and repeatedly asked M-F to switch off his light, but every request was ignored. Finally Roger's patience expired, and he took his Smith & Wesson .38 from his bedside locker and extinguished the light, which was about a foot above M-F's head! No further words were exchanged, and we all went to sleep. So far as we were aware the incident was not mentioned again, although when we cycled to breakfast that morning the occupants of the adjoining hut were observed scrutinising two unexplained bullet holes in their roof! This exploit later gave rise to a local adaptation of a popular RAF verse – "It wasn't the Almighty who put out the lightie, it was Roger the . . ." etc.'

John Bateman, gunner, 13 OTU, Bicester.

'Joyce Shaw of Harrogate and I became acquainted while I was operating out of Topcliffe. I would visit her home and parents for a meal and an overnight stay once in a while. Joyce was later to become a member of the WAAF (Women's Auxiliary Air Force). The inconvenience of having their toilet facilities in a separate brick building at the rear of their row house, located in the centre of this large city, never ceased to amaze me. This arrangement, after an evening at the pubs, left a lot to be desired . . .

'After I was posted to 76 Squadron at Middleton St George, off duty we would frequent the pubs and dance halls in Darlington and York. A few memories of these places come to mind, like the night we stole the stuffed ram's head from the Fleece, a pub in Darlington, with which to decorate our sergeants' mess. And how we had to "sheepishly" return it the following day after a complaint was made to our commanding officer . . . And, speaking of pubs, after the familiar "time gentleman please" was heard, one usually headed for the fish and chip shop. They were easily found as one could usually smell their tantalizing aroma for blocks and we were always hungry. They were most often served in a paper cone made out of newspapers, and liberally sprinkled with salt and vinegar. I still felt that good taste was due in part to the newsprint on the paper. It gave them that extra element of refinement!'

Pilot Officer J. Ralph Wood, DFC, CD, RCAF navigator.

RANKS OF THE ROYAL AIR FORCE

Air Marshal
Can leap tall buildings with a single bound,
More powerful than a steam train,
Faster than a speeding bullet,
Walks on water,
Gives policy to God.

Air Commodore
Can leap tall buildings with a running start,
More powerful than a diesel engine,
Just as fast as a speeding bullet,
Walks on water if sea is calm,
Discusses policy with God.

Group Captain
Leaps short buildings with a single bound,
More powerful than a tank engine,
Can occasionally keep up with a speeding bullet,
Walks on water in small lakes,
Talks with God.

Wing Commander
Leaps short buildings with a running start,
Is almost as powerful as a tank engine,
Is able to avoid a speeding bullet,
Walks on water in indoor swimming pools,
Talks to God if special request granted.

Squadron Leader
Can just clear a small hut,
Loses tug-of-war with tank engine,
Can deflect a speeding bullet,
Swims well,
Is occasionally addressed by God.

Flight Lieutenant
Demolishes chimney when leaping small huts,
Is run over by steam trains,
Can handle a gun,
Dog-paddles adequately,
Talks to animals.

(Punch)

"Transplant. It was originally old Freddie Bigginshaw-Hill's before he pranged!"

Flying Officer
Runs into buildings,
Recognises steam trains two times out of three,
Is not issued with guns,
Can stay afloat with a Mae West,
Talks to walls.

Pilot Officer
Falls over doorsteps,
Says: "I see no trains",
Trusted only with water pistols,
Stays on dry land,
Mumbles to himself.

Warrant Officer
Lifts tall buildings and walks under them,
Kicks steam trains off tracks,
Catches bullets in his teeth,
Freezes water with a single glance,
Because he is GOD.

MY LEAVES IN LONDON, 'THE BIG SMOKE'

'My leaves spent in London, sometimes called "The Big Smoke", presented a welcome relief from the monotonous instructional duties. Apart from visiting places in London which had great historical interest, I also visited the Beaver Club in Trafalgar square. This was a great meeting place for Canadians of all services. Then, of course, there were pubs like the Captain's Cabin and places like the Crackers Club. Theatres were plentiful, as were afternoon and evening dances in places like the Palais de Danse in Hammersmith. Young girls were plentiful at these dances – too young to work, but not too young to dance. And the Americans were now appearing in great numbers, with half-a-dozen ribbons up for making it as far as the European Theatre. One American pilot was heard to ask a barmaid for an American fly-boy drink. He explained to her that it was beer. You know, you drink one and P-38.

'Piccadilly Circus was a favourite meeting place frequented by all services and all allied countrymen. About the only civilians you could see were the prostitutes, and they were numerous. They were called the Piccadilly

'B-Blondie.' (Stanley Burgess)

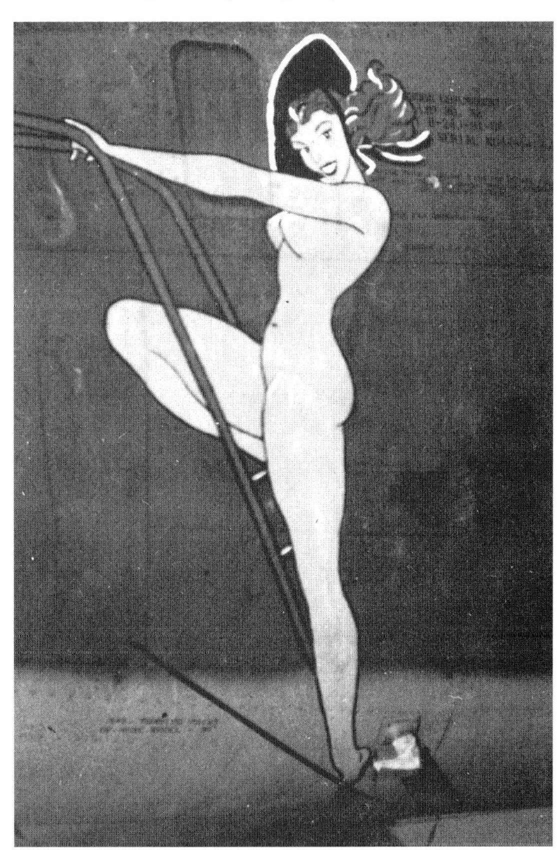

'Piccadilly Circus was a favourite meeting place frequented by all Services and all allied countrymen. About the only civilians you could see were the prostitutes, and they were numerous.' (Bill Cameron via Steve Adams)

Commandos. Just off Piccadilly Circus was the Windmill Theatre, a very popular vaudeville show due mainly to the fact that most of the show consisted of girls posing in the nude. It was all quite legal as long as they didn't move. When the pubs closed in the early afternoon we would drift over to the Crackers Club for a few hours. It was a basement bar halfway down Denman Street, a little side street starting at the Regent Palace Hotel. This was a hangout for Canadians in Bomber Command. Across the street was the Chez Moi, a hangout for Fighter Command. There was little mixing because as far as we were concerned we were fighting two different wars. Flying bombers had nothing in common with flying fighters.'

Pilot Officer J. Ralph Wood, DFC, CD, RCAF, 76 Squadron Halifax navigator.

'9 September. We have done two trips now, so I can just about call myself operational again. It's good to be back on the job again. On Tuesday night we went to Hannover. It was a very easy trip and we had no trouble at all. On Wednesday night all the squadron aircrew departed for Cambridge and we all had a good time. Thursday night we went to Karlsruhue and had very little opposition. Of course, now we have got most of France we don't spend a lot of time over enemy territory when on southern targets. As we did not get to bed until 6 o'clock yesterday morning we spent last night on camp. We went to the camp cinema and saw Bette Davis and Paul Lukas in "Watch on the Rhine", which was very good.'

Derek Smith, Mosquito navigator, letter home, 1944.

'When the two local pubs, the Red Lion in East Kirkby and the Red Lion at Revesby had beer available and we were not flying, the lads would go along and spend an enjoyable evening drinking and singing. Sometimes the Red Lion at Revesby would run out of glasses and we drank out of jam jars. I am sure we enjoyed our beer more when we were drinking from jam jars than when we had normal beer glasses. These evenings

relaxed the tension, and for a while we could forget the war and our flying missions.'
Sergeant Roland A. Hammersley, DFM, 57 Squadron Lancaster WOP-AG.

This is my story, this is my song;
I've been in this Air Force just too flaming long.

WHY WERE THEY BORN SO BEAUTIFUL?

Why were they born so beautiful?
Why were they born at all?
They're no bloody use to anyone
No bloody use at all

'Besides my instructional duties, I had two other jobs during my stay at Abingdon. One was to act as guard over two RAF officers who were under house arrest. They had just recently returned from the Middle East and their crime was getting caught in the station chapel, making love to a couple of WAAFs.'
Pilot Officer J. Ralph Wood, DFC, CD, RCAF navigator.

My Mistress
My mistress is the sky,
She calls forever "Fly".
Often she treated me shamefully —
Tried putting me to the core
Yet still I cry for more
And yearn for her embrace,
Her clear and open face
Above the cloud and rain
O let me fly again.

Jasper Miles

How is the Met, sir, how is the Met, sir,
How is the Met? – it looks very dud to me.
Let's scrub it out, sir, let's scrub it out, sir,
'Cos I've got a date fixed with my popsie.

'You could drink all the beer you wanted on a station or at the village pubs and there were always plenty of WAAFs for company.'
Piece of Cake *by Geoff Taylor, RAAF Lancaster pilot.*

'I made a number of friends at Swinderby, like the WAAFs in the various stores, the parachute stores in particular, a place visited every time we flew. All parachutes were returned there after each flight for checking and storing, being reissued as and when required. The pace of our training was such that, apart from a few quiet moments with the ladies and the occasional visit to Lincoln, it was a case of all work and very little play!'
Sergeant Roland A. Hammersley, DFM, 57 Squadron Lancaster gunner.

'Pre-war, Jock worked for Montague Burtons tailors. He was a Scot with a bizarre sense of

'There were always plenty of WAAFs for company.' (Gordon Kinsey)

humour. He considered that anything on camp that was left alone or unattended could be "borrowed" or "moved". Apparently he was having an affair with an attractive lady school ma'am in Tyler, Texas. He was certainly besotted with her. About 48 hours before we were to depart for New York, Jock vanished. So did one of the Stearmans from the line. They were all fully fuelled.

'I learned later that he had cranked up the kite on his own, had taken off and flown to Tyler, where he landed in a small field straight into the arms of his lady love. The Stearman could not be flown out of the field. It was too small! It had to be disassembled and hauled out. Some months later, while I was in a camp hospital near Blackpool, Jock looked me up. He was eventually picked up by the local police and handed over to the RAF MPs, brought back, found guilty of going AWOL and got one month inside. He then remustered. I have never seen or heard of him since.'

L. James Freeman, cadet pilot, No. 1 BFTS, Terrell, Texas.

'One of the favourite places for spending a while with a favourite girl was in the boiler house on our accommodation site. It was nice and warm, the boiler man left a seat there with just enough space on it for two. By now I had friends among the girls serving with the WAAF, and I found that with care I could make it to the boiler house before any of my colleagues. They also found it a warm place for spending an hour or so cuddling and kissing out of sight and, I hope, away from prying eyes.'

Sergeant Roland A. Hammersley, DFM.

'The average bomber station was devoid of news. We seldom saw a newspaper, and occasionally heard the BBC news if we happened to be in the mess when the wireless was on. I was on leave in

'Sometimes we lost a lot of friends, and you could see WAAF red-eyed from weeping for one special boy . . .' Flying Officer Henri Cabolet, Section Officer Jean Barclay, and Flt Lt R.V. Smith, 239 Squadron, West Raynham, 1944. (via Tom Cushing)

August 1943 and so happened to meet a young female in whom I was interested. We didn't talk about what we'd do when the war was over; it was never going to end. We lived for the next leave in six weeks' time – we had double leave because flying was considered to be more dangerous than staying on the ground. Oh yes, they gave us an extra bob a day as well – flying pay, not "danger money", as the more daring participants would have it.'
Geoff Parnell, air gunner.

'Nelly was a Titian-haired WAAF.

'Mainly, I wanted to see her about collecting the crew's flying rations, which were our due when we were operating. Made up of chocolate, boiled sweets, chewing-gum, and canned tomato juice, most of it, except for the gum and tomato juice, found its way to the kids in the households of our English crew members.

'Waiting for the arrival of the bar sergeant to issue the rations, I made a date with Nelly for the following night. "I'll meet you at the Blue Boy in the village at seven," I said.

'Nelly looked at me with a pair of quizzical blue eyes. "Don't be late," she said.

'We left it at that.'
Geoff Taylor, RAAF, Lancaster pilot, author of Piece of Cake *(George Mann, 1956).*

'I remember hearing the bombers leave and counting them coming in over the field early next day. Sometimes we lost a lot of friends and you could see WAAFs red-eyed from weeping for one special boy. A terrible loss of human life, but you had to carry on with what had to be done.'
Daphne Smith, Women In Air Force Blue *(PSL).*

'Was advised that if we came down in the desert and were rescued by Arabs, the women would act as sort of servants, but on no account should we thank them or even smile at them.'
M. Howland, RAAF, pilot.

2ND TAF

We're flying binding Bostons
At 250 binding feet,
Doing night intruders
Just to see who we might meet.
And when the daylight dawns again
And when we can take a peek,
We find we've made our landfall
Up the Clacton binding Creek.

'In the summer of 1943, 2 Group was transferred from Bomber Command to the newly formed 2nd TAF, and later that summer, when we were at Hartford Bridge, a captain ALO (Army Liaison Officer) gave us a talk in his experience with 1 TAF, and finished by reciting "The Flying Brothel":

Gestapo spies among our forces
Investigate subversive sources
Whereby Teutons sensing spring
Feel the urge to have their fling.

Wilhelmstrasse finds of late
A tendency to masturbate,
And with their natural superfluity
Of good Germanic ingenuity
Seek a suitable diversion
To put an end to this perversion.

A Flying Brothel is the thing
To ease the urgencies of spring
For then each soldier dissolute
May have his airborne prostitute.

Berlin was scoured, and Dresden, too
To find sufficient trollops who
Could give the joys of copulation
Yet be au fait with aviation.

". . . approaching the target, showers of . . . er . . . stuff came up . . . we jinked like . . . er . . . anything, which completely . . er . . . messed up our bombing run . . . we turned and dropped the whole bl . . . inking load through a hole in the flak!"

At last the scheme has reached fruitiion
Our troops are in top-line condition,
And though their ersatz food may cloy
They still can have their "strength through joy".

'We're flying binding Bostons at 250 binding feet . . . (RAF Swanton Morley)

'Within a minute we were over the Dutch coast, flying over flat silt land, when all hell was suddenly let loose. Without warning the whole flak battery must have opened up, and the sky around us was full of flak bursts. They had got our altitude absolutely spot-on. It was just like going through a wall, and how we escaped must remain one of life's mysteries. I saw the four aircraft on our left take the full punishment. Dickie's 'plane took a direct hit in the starboard engine, which fell from the nacelle and rolled

The breakout from the beachheads and the advance across France and the Low Countries were assisted by thousands of sorties by Spitfires, Tempests, Typhoons and Mustangs (pictured), which were called upon whenever resistance was encounterd. (Harry Holmes)

On D-Day no fewer than 16 Squadrons of 2nd TAF were equipped with Typhoon 1a/1b fighter-bombers. (BAe)

along the ground like a gigantic ball of fire. His aircraft pitched sideways, cartwheeling into the ground at 280 mph. The other three must have been hit almost instantaneously and all hit the ground in a complete shambles of fire, smoke, and scattered pieces of metal. I couldn't see exactly where the flak was coming from and just fired my guns in the general direction without taking any sort of aim. Johnny had his finger on the button and exhausted nearly all his ammo. What seemed like an eternity probably lasted only a couple of minutes or so and then we were out of range. I could see the following aircraft getting a fairish dose, but they were jinking all over the sky and not flying straight into the stuff as we had done. I saw only two of them go down. This was my first trip and it was the only time in all my operational experience that I had seen such a large formation broken up so completely.'

Pilot Officer John Bateman, Boston rear gunner, 107 Squadron, 22 October 1943.

On D-Day the RAF flew a total of 5,656 sorties with only very light losses. The break-out from the beachheads and the advance across France and the Low Countries were assisted by thousands of sorties by Spitfires, Tempests, Typhoons, and Mustangs, which were called upon whenever resistance was encountered. The RAF's contribution to the Allied Expeditionary Air Force was the 2nd Tactical Air Force (TAF), which by 1945 consisted of four groups with a total of 81 squadrons. Units of the 2nd TAF began to move to forward airfields in France just two weeks after D-Day, and by the end of the war many squadrons were based on abandoned Luftwaffe airfields in Germany itself.

Once the Camel made them dance
O'er the fields of Northern France
As its forebears called the tune
So (with knobs on) does Typhoon

On D-Day, no fewer than 16 squadrons of 2nd TAF were equipped with Hawker Typhoon 1a/1b fighter-bombers.

'When Basil Embry was AOC 2 Group he would regularly visit the stations under his command. He was quite small and had the most piercing blue eyes – on first acquaintance he could appear to be quite forbidding, particularly if things were not going according to his liking, but he was a

sociable man and on both occasions when he flew with us, and after debriefing, he insisted that we all go back to the Mess with him for a drink.'
John Bateman.

Mosquito FB.VI squadrons in 2nd TAF also intruded over the Reich, bombing and strafing German lines of communication and Luftwaffe airfields, but the aircraft is probably best remembered for daylight precision operations, particularly the pinpoint raids on Gestapo buildings in occupied Europe which made it famous in 1943–45. Mosquitos served in a multitude of roles including nightfighting, bombing, pathfinding, and photographic reconnaissance.

MOSSIES

Mossies they don't worry me,
Mossies they don't worry me,
If you get jumped by a One-nine-O,
I'll show you how to get free.
Keep cool and collected,
Keep calm and sedate,
Don't let your British blood boil.
Don't hesitate,
Just go right through the gate,
And drown the poor bastard in oil!

'The pilot and navigator sat side by side in this "Wooden Wonder", or "Termite's Delight", as it was sometimes called. The pilots had a steel plate under their seats to protect them. Navigators had an extra sheet of plywood. We all had a nagging fear that our jewels might be shot off. The moral seemed to be that pilots make better fathers.'
Pilot Officer J. Ralph Wood DFC, CD, RCAF, 692 Squadron Mosquito navigator.

Grace

Her name was Grace, she was one of the best,
But that was the night I gave her the test.
I looked at her with joy and delight
For she was mine, and mine for the night.
She looked so pretty, so sweet and slim,
And the night was dark, the light was dim.
I was so excited my heart missed a beat,
For I knew that night I was in for a treat!
I had seen her stripped, I had seen her bare,
I felt her round, and felt her everywhere,
But that was the night I liked the best,
And if you wait I'll tell you the rest.
I got inside her, she screamed with joy.
For this was her first night out with a boy.
I got up high, and quick as I could,
I handled her well for she was good.
I turned her over upon her side,
Then on her back – that was all tried:
I pushed it forward, I pulled it back,
Then I let it go, until I thought she would crack.
She was one great thrill, the best in the land,
The twin-engined MOSQUITO of Bomber Command.

Sergeant Harry Tagg, 1655 Mosquito Training Unit, RAF Marham, April 1943 (via Daphne Light, ex-Marham WAAF).

'A month at the Mosquito Training Unit and we headed for our new station, 692 Squadron RAF Graveley, near Cambridge in eastern England. We shared this station with 35 (Pathfinder) Squadron, which was flying Lancasters. A friendly rivalry existed between these two squadrons, especially when we were both frequenting the same local pub or the officers' mess. While our Mosquitoes roamed the German skies in all kinds of weather, the heavies (mostly Lancasters and Hallybags) were more particular about when they went aloft. We took special delight in provoking the gentlemen who flew the heavies by singing our song "We Fly Alone" in the pubs we both frequented. Our rewritten lyric of the juke-box favourite "I'll Walk Alone".'
Pilot Officer J. Ralph Wood DFC, CD, RCAF 692 Squadron Mosquito navigator.

We fly alone, when all the heavies
are grounded and dining
692 will be climbing.
We still press on

It's every night, but though they
never can give us a French route
To the honour of 8 Group
We still press on.

It's always the Reich, no matter how far
One bomb is slung beneath
The crew they are twitching
It's not fear of ditching
It's twelve degrees east —
One engine at least!

We fly alone, although tonight
you can see it's a stand-down
But tomorrow the big town
We still press on.

'One unusual return from enemy territory, and most satisfying, included a dive beginning at the French coast from 32,000 ft to 10,000 ft, reached at Southwold on the English coast. This 88-mile journey was completed in 11 min, which was fast, even for a Mosquito. With our cookie (4,000 lb bomb) gone, our two 50 gal drop tanks discarded, and our fuel load pretty well depleted, it wasn't too hard to accomplish this feat.'
Pilot Officer J. Ralph Wood DFC, CD, RCAF 692 Squadron Mosquito navigator.

'24 August. Well, I have been flying in a Mossie at last. They are wizard aircraft and Phil is a very good pilot. There isn't a lot of room after a Lanc though. Yesterday afternoon we did two hours stooging around. This morning we did a cross-country which took us out over the Isle of Man and the north of England. It was wizard up at 25,000 ft; at one time we could see England, Wales, Scotland and Ireland. After flying we had the rest of the day off, but as it has rained ever since it wasn't much good. What lousy weather! Anyone would think it was November instead of August.'
Derek Smith, Mosquito navigator.

Top-scoring V1 Killers

1 Tempests of Newchurch Wing 357

2 ditto 223

3 Spitfire XIVs 185

4 Mosquito XIIIs of 96 Sqn 174
(Ford)

6 Mosquitos of 418 Sqn 90
(Holmesby, Hurn & Middle Wallop)

From Doodlebugs & Rockets *by Bob Ogley (Froglets 1992).*

Almost Home

Single engine? Keep on turning!
'Tis our gratitude you're earning,
For we nearly 'copped our lot'
And our undercart is shot.
Dear old Mossie? Wooden steed?
One more mile is all we need!

Jasper Miles

'On 20 January 1944 Fg Off Hugh Cawker and I were briefed for a target east of Toulouse. We were airborne on a Mosquito IX at 11:20 and climbed to 10,000 ft over Benson before setting course for Toulouse. We crossed out at Selsey Bill, on track, at 23,000 ft, and found contrails down to 23,500 ft. We debated whether to turn back, but hoping trails would rise over the Channel we continued, and finally crossed in at Caen at 25,000 ft, just under trails.

'We flew over broken cloud in the middle of France, but the sky was practically clear at Toulouse. We did one run over Toulouse and Toulouse-Blagnac airfield for luck, and then set course for the first target. After covering the first two targets I noticed oil escaping from the starboard engine (caused by a split in the "banjo" union (oil pipe), but as the temperatures and pressures seemed normal, and as the remaining targets were in the general direction of home, we decided to continue. However, while doing the sixth target the oil pressure began to fall. I called Hugh back from the nose and he turned on the emergency oil supply. However, this did not seem to help and at 14:00 hrs, when the oil pressure was nearing 25 lb, I feathered the starboard airscrew.

'We first thought of going to Corsica, but we decided there was too much water to fly over and we were not sure where the aerodromes were, anyway. We next considered returning to base, but I knew there was a lot of activity on the north French coast that day and I did not like the idea of coming out through it at 10,000 ft on one engine. Besides, if anything did happen on the way it would mean walking all the way back again.

'Finally we decided to set course for Gibraltar. I did not know what our petrol consumption would be on one engine, and I did not think we had enough to reach Gib, but we figured we would fly as far as possible, then bale out and walk the rest of the way.

'We set course for the east coast of Spain and decided to fly down it. I gradually reduced height from 26,000 ft to 12,500 ft, where we were able to maintain height without overheating the port engine. We had maps of Gib and Marseilles but nothing in between. However, after flying some time I remembered I had a silk escape map hidden in the lining of my trousers. I had to practically undress to get at it, but with it we were able to get the outline of the coast and the approximate distance to Gib.

'By this time we had also figured out our petrol consumption, and we figured that there was just a faint chance of making Gib as long as the engine held out. We set an approximate course inland for Gib, and I decided to risk using some of our precious current (the generator is on the starboard engine) on sending out an SOS in the hope of getting a vector to Gib. So we called both on W/T and VHF, but received no reply.

'We found it impossible, as soon as we left the coast, to map-read on my small map, so Hugh busied himself getting ready to bale out. We tore off the inner hatch, tested his harness and parachute, etc. Remembering an old escape lecture which told us to get rid of all foreign money when crossing into Spain, we opened our money packets, kept the escape aids and put the money in the navigation bag to be destroyed with the aircraft.

'After two hours on one engine we were getting a bit tired, but we had computed our petrol consumption again and it proved to be less than the first estimate. If only that engine would hold out!

'After three hours we found ourselves over the Pyrenees mountains with approximately one hour left. We now figured we must be on our Gib map (which included the south coast of Spain), but we realised by this time we must be near the south coast. We flew over the mountains full of expectancy, and we pinpointed ourselves on the coast at Malaga. There, silhouetted against the sinking sun, was the Rock of Gibraltar.

'We circled the Rock at 10,000 ft, descended to 2,000 ft over the runway, and fired off the colours of the day. I then made my first single-engined landing in a Mosquito after spending six and a half hours in the air, three and a half hours of which were on one engine over the Pyrenees, and it did not even heat up.'

Flight Lieutenant John R. Myles RCAF, 544 Squadron PRU Mosquito IX pilot.

SPOOFING

As I was walking up the stair
I met a man who wasn't there.
He wasn't there again today.
I wish, I wish he'd stay away.

Hughes Mearns

'Our aircraft would accompany the main bombers stream and then circle above the target; the special operators used their transmitters, in particular* Jostle*, to jam the German radar defences . . .' (Jerry Scutts)

On 23 November 1943 100 Group (Special Duties – later Bomber Support) was formed under the command of Air Cdre (later AVM) E.B. Addison to bring together exisiting radio countermeasures (RCM) and Electonic Intelligence (ELINT) operations and help significantly to reduce Bomber Command aircrew casualties. Stirlings, Wellingtons, Halifaxes, Liberators, and Fortresses variously equipped with 32 different types of RCM, in conjunction with ground RCM equipment, made feint attacks on the enemy heartland, while Mosquitos gave direct support to night bombing and other operations by attacking enemy nightfighters in the air and over his bases. It was a clandestine war of move and counter-move in which first one side, then the other, attempted to render the opposition's radar and homers ineffective before new countermeasures re-established ascendancy again.

'Operations were of two distinct types. In the first, two or three of our aircraft would accompany the main bomber stream and then circle above the target; the special operators used their transmitters, in particular *Jostle*, to jam the German radar defences while the Lancasters and Halifaxes unloaded their bombs. Then everyone headed for home. The majority of our operations were the second type; *Window* spoofs. The object of these was to confuse the enemy as to the intended target. There was a radar screen created by other aircraft patrolling in a line roughly north to south over the North Sea and France. A group of us, perhaps eight aircraft, would emerge through this screen scattering *Window* to give the impression to the German radar operators that a large bomber force was heading for, say, Hamburg. Then, when the Germans were concentrating their nightfighters in that area, the real bomber force would appear through the screen and bomb a totally different target, perhaps Düsseldorf. After several nights, when the Germans had become used to regarding the first group of aircraft as a dummy raid, the drill was reversed; the genuine bombers would appear first and with luck be ignored by the German defences, who would instead concentrate on the second bunch, which was of course our *Window* spoof. So we rang the changes, sometimes going in first, sometimes last, in an attempt to cause maximum confusion to the enemy, dissipation of his resources, and reduction in our own bomber losses.'
Sergeant Don Prutton, B-24 Liberator flight engineer, 214 Squadron, 100 Group.

'There was no shortage of nightfighter aircraft. From the middle of 1944 onwards we could even speak of a surplus. The decrease of the German nightfighter successes in this period was mainly due to interference, shortage of fuel, and the activities of 100 Group. The task of this specialist unit was to mislead our fighters and to befog our conception of the air situation by clever deceptive manoeuvres. This specialist unit finally solved its task so well that it was hardly ever absent from any of the British night operations, and it can claim to have set really difficult problems for the German nightfighter command. The British increased their raids at the end of 1944 from month to month, with decreasing losses.'
Adolf Galland.

THE BIG CITY

'Tonight you are going to the Big City. You will have the opportunity to light a fire in the belly of the enemy that will burn his black heart out.'
Air Marshal Arthur Harris, C-in-C, RAF Bomber Command.

Thirty-five major attacks were made on Berlin and other German towns during the Battle of Berlin, between mid-1943 and March 1944; some 20,224 sorties, 9,111 of which were to the Big City. From these sorties (14,652 by Lancasters), some 1,047 aircraft failed to return, and 1,682 received varying degrees of damage.

'. . . I felt quite elated. We had actually bombed the capital of Germany. But the trip wasn't that pleasant. I thought about that goddam "Butcher Harris" (Bert). Butcher was the deserved nickname of the RAF Chief of Bomber Command. He didn't give a damn how many men he lost as long as he was pounding the shit out of the Germans. He was just as willing to sacrifice Englishmen as Canadians.'
Canadian navigator.

'We can wreck Berlin from end to end if the USAAF will come in on it. It will cost between

Thirty-five major attacks were made on Berlin and other German towns during the Battle of Berlin between mid-1943 and March 1944 . . . 14,652 by Lancasters . . .' 'We can wreck Berlin from end to end . . . if the USAAF will come in on it. It will cost between 400–500 aircraft. It will cost Germany the war.' Air Marshal Arthur Harris. (via Tom Cushing)

400–500 aircraft. It will cost Germany the war.'
Air Marshal Arthur Harris.

And so it's Berlin or Bust
Oh, we didn't want to do it, but we must, boys.
Berlin or Bust,
Someone started kicking up a fuss, boys.
Tell them the truth we loudly roar,
And so the RAF have posted leaflets through the door.
So it's Berlin or Bust.
Oh, we didn't want to do it but we must"

'Op No. 29, Berlin, 7 July 1944. We're going right to the Snake's home this time – *Berlin* – quite keyed-up too, but not letting on to each other. The trip was successful, but packed with excitement. As we did our bombing run into the centre of Berlin our starboard engine seemed to

catch fire. Andy feathered it immediately in case of fire spreading. We finished our run, dropped our Cookie, and were immediately coned by a great number of searchlights. After five minutes we got out without damage. On the return journey we were coned and shot at again in the Hamburg district. Finally, we reached the English coast and landed at the nearest aerodrome with about 20 gallons of petrol left. We returned to base the following day after our engine was repaired. We have now been over the hottest target in Germany and feel quite good about it. One of our crews failed to return.

'Op No. 30, Berlin, 10 July '44. Well, we are on the "milk run" again – *Berlin* – or bust. Of course, we had to get coned and shot at again over the target. As if that wasn't enough, the boys of Heligoland had a crack at us too. The trip was exciting and a good one, but we lost a very fine crew. The pilot was our CO, a New Zealander. They think he got it near Cologne. Andy and I saw a lot of action in that direction as we passed it. On return we just made the English coast and landed with 5 gallons of petrol left. We are getting along fine together and our teamwork is improving. Andy likes the choice of names I use for the Hun when he lets us have a barrage of flak. We are still keen and a bit more Berlin minded. We were coned for 9 minutes there this time.

'. . . Op No. 77, Berlin, 3 November '44. This, my last operational trip, is one that I've wanted to write about for quite a while. I could write pages about our feelings, our twitch, and our relief when we got back again; but to be brief, we went to none other than our old favourite (?) – *Berlin*. Our 17th trip there and I hope our last. We took off at midnight. I've never felt so keenly about a trip in all my life. I certainly *lived* every moment of it. There were numerous fighter flares and fighter contrails all the way there and back. Saw one jet fighter. The moon was rather bright, too. The raid was wizard, though. Andy and I exchanged congratulations and shook hands on it before we got out of the kite. Gosh, we were a couple of happy kids. *Moncton Express III* came through with flying colours. And so endeth my second and last tour.

PS. Andy's a damn good driver.'

Pilot Officer J. Ralph Wood, RCAF Mosquito navigator, 692 Squadron.

'15 October. Since I last wrote we have done two trips. On Wednesday night, or should I say

Thursday morning, we were on the "milk run" to the City. It was an easy trip but we took off at 1.45 and landed at 6.30 so we were rather tired. Take-off was rather shaky as we had a crosswind of 30 kt. We thought we should be doing our 13th trip on Friday the 13th, but as we did not take off until this morning it was the 14th . . .

'2 November. Last night we went to the Big City. My 20th trip this tour and it completed my half-century. It was a wizard trip with a full moon and we saw the Keil Canal, Lubeck bay, and all stations east quite clearly. We were very sure it was the City because we could see the streets and the bridges over the river. The opposition was far from City standard. It was a long route and took us 5 hrs 20 min, which is a long time in a Mossie.

'3 January 1945. We have done three trips in the last three nights to Berlin, Hannover, and Nurnburg. It's a nice way of spending New Year's Eve, going to Berlin. One thing, there was a party in our mess, so when we got back we had a wizard meal.'

Derek Smith, Mosquito navigator, in letters home, 1945.

'24 March 1944, a memorable day. On arrival at the Flight Office we found our names on the battle order for the night's operations. The aircraft we were to fly was ND405 *T-Tommy.* We set off on bicycles that had been issued to each one of us, to look the aircraft over and check the equipment. The ground crew responsible for the maintenance of ND405 "T" were a fine bunch and gave us much information as was possible about it as we went through the checking procedure. The bomb load was 1 x 4,000 lb, 48 x 30 lb and 600 x 4 lb.

'Later we were fully briefed both as individual crew members and then all crews together. We soon learned that the target was Berlin – the Big City. At the briefing we were told at what time there would be signals broadcast from Bomber Command; when we would receive weather reports; where the searchlight belt and anti-aircraft guns were known to be, and also the positions of known German nightfighter units and airfields en route. A weather report was given by the Station Met Officer, the indications being that the weather conditions were not too good and we would be meeting quite strong winds at 18,000 to 20,000 ft. We were issued with amphetamine (wakey-wakey) tablets; these were taken just prior to take-off and would keep the crews wide awake and on a "high" for the duration of the flight. If the operation was cancelled, it meant a sleepless night which, for the most of the crews, meant that a wild night of drinking would take place in both the Officers' and Sergeants' mess until the effects of the drug wore off and sleep could take over.

'It was customary for a meal to be prepared for the crews before we flew. We were then issued with a flask of tea or coffee, with chocolate, sandwiches, and an apple; armed with a 0.38 revolver and parachute. Codes and Very pistol with cartridges which when fired would give the coded colours of the day. We were even given what were understood to be those in use by the German forces that day. After emptying my pockets and locking my personal items into my cage-type locker, I joined the crew in the crew bus with WAAF Connie Mills at the wheel. The bus that collected the crews from near the control tower was often driven by Connie. We were then taken out to ND405 "T". We had another look around the aircraft with the ground crew and about an hour before we were due to take off we settled into our places to await the take-off order. When the first part of the take off procedure commenced, we were lined up on the airfield perimeter with 17 other Lancasters from the squadron. All crews would by now have taken their amphetamines and would be wide awake.

'The first Lancaster was given the "green light" from the mobile watchtower, and we watched as it slowly climbed away. The remainder all slowly moved around the perimeter track towards the runway and then it was their turn for destination Berlin! The smoke from the engines and the smell of burning high-octane fuel eddied across the airfield. Some 60 tons of explosives and incendiaries were to be dropped by 57 Squadron that night, and the sight of 17 Lancasters, each under full throttle, roaring away into the evening sky, was an awesome spectacle. Sergeants Frank Beasley and Leslie Wakerell with their ground crews and a number of other well-wishers watched us away before returning to while away the long hours before our return. The smoke and smell slowly thinned and drifted away over the silent airfield, and we were on our way to our first bombing operation with the squadron.

'We were airborne at 18:45 hrs. This was to be the order of things for some time to come. As the weather reports came in and I decoded them, it became apparent from Mack's findings that they were not as he expected them. We were faced with greater wind speeds than those indicated in the signals being sent out to us from Command, so we used our own. We were late arriving over the target, and we could see there were great fires as the run-in towards the target commenced.

'Having bombed successfully, we headed back towards home, only to be told that we would have to land away, and so we spent the night at RAF Coltishall, an airfield in Norfolk where fighter squadrons were based. The time we spent flying was 7 hrs 30 min. We were debriefed and fed, then shown to our sleeping quarters. We made the 35-minute flight back to East Kirkby the following afternoon, leaving at 15:00 hrs, by which time the fog that had prevented our landing the previous night had cleared. Of the 17 Lancasters from the squadron that flew this operation, one made an early return and two others failed to return. We made our reports at the squadron office before heading for our huts to await the evening meal.'

Sergeant Roland A. Hammersley DFM, WOP/AG, 57 Squadron.

'The trip to Berlin on 24 March stands out in my memory due to the fact that the forecast winds were far from what we actually encountered. I suspect that we probably never came near to Berlin and, on the way home, found ourselves far to the south of where we were supposed to be. Directly ahead of us was the Ruhr Valley with its heavy defences, and we elected to head to the north and pass between Hamburg and Hannover on our way to the North Sea. We were lucky enough to get to the coast without any trouble, but by now we were running considerably late because of the extra distance we had flown. Fuel became a major concern as we crossed the North Sea, and at one point my flight engineer, when asked, reported that all gauges were reading zero but, according to his calculations, we had about thirty minutes' fuel left! His calculations proved correct, and we landed at Coltishall, Norfolk, whereupon all four engines quit and we had to be towed off the runway!'

Larry Melling, pilot, 51 Squadron, Snaith, Yorkshire.

'Suddenly, on this night, the utter blackness of the barrack is flooded with white light from the windows as if a desert high noon had come without warning to blacked-out Saxony.

'"Fighter flares," comes the shout. "The Jerries are dropping them right across the stream."

'High above the camp the clouds are glowing with five great bright nuclei. A corridor of white light marching across the bombers' route. They are Luftwaffe fighter flares all right. Everybody in the barrack has good cause to recognize them. Silhouetted against the flares below them, the bombers run the gauntlet with the German nightfighters waiting above them.

'Faintly, above the tumult of engine noise in the berserk sky, comes a faint, vicious flurry of machine-gun and cannon fire.

'In the few minutes it takes the hanging flares to drift lower and burn out, five British heavy bombers have screamed down around the camp, hurtling into oblivion with their bomb loads. Several are close, and the blast from the thundering explosions sways the barrack as if it were a tent.

'Rolling someone off your back, you stand up on the brick floor, which had felt like a bed of roses when you dived for it, and peer out of the window.

'In the darkness to the west, five bright fires are burning on the ground, and even as you watch you can see the initial yellow glow of the incendiaries melting into white as they burn.

'"Bang, bang, bang, bang, bang," comments a voice alongside you. "Those poor bastards."

'Five less Lancasters for Berlin. Five new crews reporting to their squadrons this week. Five unrecorded minutes of action in yet another air assault on Berlin. Iron Crosses. Wooden crosses. The luck of the game.'

Geoff Taylor RAAF, PoW, Stalag IVb, Muhlberg-on-Elbe, Saxony, in Piece of Cake *(George Mann, 1956).*

PATHFINDERS

Per Ardua Ad Astro

To the memory of the splendid Personnel
of No. 7 Squadron RAF Pathfinders
Oakington, England, WWII 1943–1945
'With whom I had the honour to serve.'

The Phantom Observer of No. 7.

'We shall bomb Germany by day as well as by night in ever-increasing measure, casting upon them month by month a heavier discharge of bombs, and making the German people taste and gulp each month a sharper dose of the miseries they have showered upon mankind.'
Winston Churchill.

Number 8 Pathfinder Force (PFF) Group was formed on 15 August 1942 within Bomber Command under the dynamic leadership of Gp Capt Don C.T. Bennett, RAAF. Its task was to 'find' (using *Gee*, and later *H2S* and *Oboe*), 'illuminate' and 'mark' targets for the main force. Initially five squadrons – Nos 7 (Stirlings), 35 (Halifaxes), 83 (Lancasters), 156, and 109 Squadrons (Wellingtons) were used. By March 1945 8 Group numbered 19 squadrons, 11 of them equipped with *Oboe* Mosquitos.

'Amongst the many vistors to Pathfinder Headquarters we had on one occasion an American lady of some distinction, Mrs Ogden Reid of the *New York Herald Tribune* . . . I took her to see a Mosquito squadron take off. We stood at the end of the runway, and the Mosquitos roared off in quick succession close behind each other . . . As the last aircraft left she said: "I won't ask what their destination is, but I can guess that it is the usual milk run to Berlin. Tell me, what is their bomb load?"

'I said: "The Mosquito bombers carry a 4,000 lb blockbuster to Berlin."

'She thought a little, and then she asked: "And what do the B-17 Flying Fortresses carry to Berlin?"

'I replied, "At present, with the routeing which they used and with the larger load of ammunition necessary for daylight operations, they are carrying 3,500 lb. In any case, they cannot carry a blockbuster, as it is too large for their bomb bays."

'Mrs Ogden Reid looked very solemn. "I only hope," she said, "that the American public never realises those facts."'
AVM D.C.T. Bennett CB, CBE, DSO, Pathfinder *(Panther 1960).*

'The Mossie could carry as big a bomb load to Berlin as the US Flying Fortress, which needed an eleven-man crew. The "Light Night Striking Force" of Mosquitos raided Berlin 170 times, 36 of these on consecutive nights. The Mosquito flew so often to Berlin that its raids were known as the Berlin Express, and the different routes there and back as platform one, two, and three.'
Pilot Officer J. Ralph Wood DFC, CD, RCAF, 692 Squadron, who flew 50 ops as a Mosquito navigator, 6 July–5 November 1944.

'12 May 1943. *Ops Duisburg* K. Paramatta. Easy and very quiet trip. Bombing concentrated. 572 on. 5.9 per cent lost. 4.05 hrs – our shortest trip.

13 May 1943. *Ops Skoda Works Pilsen* K. Newhaven. From our point of view it was a very successful trip as we got an aiming-point photograph of Skoda. It was a very long and boring trip . . .

23 May 1943. *Ops Dortmund* K. Paramatta. Easy trip. First back. Photograph just North West of aiming point. Devastating attack. 826 on. 1.6 per cent lost. 4.30 hrs.

25 May 1943. *Ops Dusseldorf* K. Paramatta. Nearly finished where we started. No problems except with cloud. Bombing very scattered. 759 on. 3.6 per cent lost. 4.30 hrs.

27 May 1943. *Ops Essen* K. Wanganui. 10/10 all the way, as it should be for a last trip, so we did it! 518 on. 4.4 per cent lost. 5.10 hrs.'

Sergeant Derek Smith, navigator of 'K-King', 61 (Lancaster) Squadron, Bomber Command, 5 Group, Syerston.

'. . . So far as the Main Force bombers were concerned, we detailed the method [of marking] by three codenames. These were "Newhaven" for ground marking by visual methods, when the crews simply aimed at the target indicators on the ground; "Paramatta" was when we ground-marked, using *H_2S* only, owing to bad visibility or broken cloud; and finally "Wanganui", which was pure skymarking, when Main Force crews were required to bomb through these sky-markers on a required course which we detailed . . . The names . . . were chosen very simply. I asked one of my air staff officers, Sqn Ldr Ashworth: "Pedro, where do you come from?" And he replied: "From Wanganui". I then said: "Just to keep the balance with New Zealand, we will call the blind ground marking by the name of Parramatta". Then, looking for a third name for visual ground marking, I pressed the bell on on my desk and summoned Cpl Ralph, my confidential WAAF clerk. When she came in I said: "Sunshine, where do you live?" And she

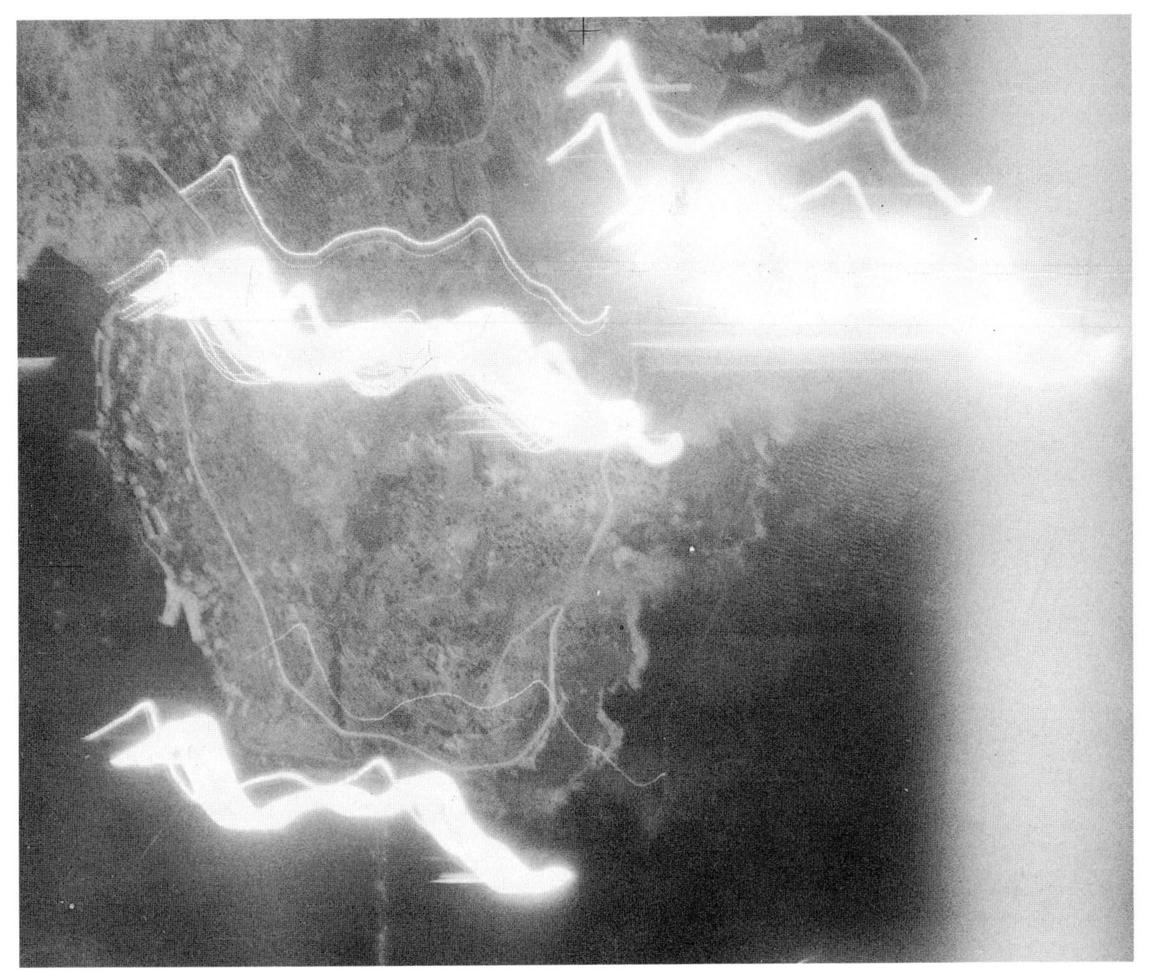

'So far as the Main Force bombers were concerned, we detailed the method [of marking] by three codenames.' (Flt Sgt Walker)

replied: "Newhaven". Thus it was that these famous codenames were born.'
AVM D.C.T. Bennett CB, CBE, DSO, Pathfinder *(Panther, 1960).*

'You have just walked away from a window, eyes stinging from staring out curiously towards Magdeburg, when there is a shattering explosion above our very roof, it seems; an explosion which is all the more shocking because of its unexpectedness, its detachment from the distant raid.

'There is a yell of sheer terror from one of the windows. "Jesus Christ! TIs right over the camp!"

'Sheer animal panic sweeps the barrack.

'In the language of Bomber Command "TIs" means "target indicators" – the vivid flares with which the Pathfinder Force marks the aiming point of a target, the glowing bullseyes on which the main force stream concentrates its bombing.

'In the darkness there is a headlong rush to get out of the barrack. The wooden doorway creaks with the sudden impact of hurtling bodies jammed against each other. You don't join them. If the Pathfinders have blundered, and in a few seconds or minutes the stalag is to be the focal point for a misdirected mass raid, then you'll die anyway, either by bombing or the guns of the guards if you run.

'Outside, the low-hanging sea of clouds is stained an eerie green by the hanging flares on which, as we wait, hundreds of bomb sights may be lining up.

'You are scared, more scared than you have ever been before in your life. There is nothing, absolutely nothing, you can do about it. Suddenly you would like to vomit.

'The initial panic at the doorway has ended, and men are just standing at the windows, lying on the brick floor, unthinkingly obeying their individual reactions and instincts.

'Then, abruptly, the green flares have burned out and the whine of an aircraft's engines is blending away westward into the rumbling of the bombing at Magdeburg.

'Momentarily there is complete and utter silence in the barrack, until a relieved, youthful voice pipes up: "Hell, there was nothing to worry about."

'"Sure, agrees a gravel-voiced Canadian, "but next time you walk all over my face, TAKE YOUR GODDAM BOOTS OFF."

'Next morning you are yarning with a South African in the army compound. "Man!" he says enthusiastically, "did you see those green flares last night? The prettiest things I've ever seen. I could have watched them all night."

'You start to enlighten him and then stop. He's happier that way.'

Geoff Taylor, Stalag IVb. Piece of Cake *(George Mann, 1956).*

Cheerio Chaps

I watched their calm and quiet wait
As nightfall beckoned their coming fate
I watched them when they took to wing
With flashing smile and "thumbs up" grin

I watched their stunning youth cut short.
"Press on," they cried, "We can't abort,
For who will end this Bleeding War?
Our lives are pledged – we can't do more."

I watched them brave the flames of Hell
I watched their pain when wingmates fell.
I watched them plunge their way to doom,
A flaming Lanc – their final tomb.

Who were these men I watched with care?
They saw me not, but I was there.
Brave, simple chaps who knew stark fear,
"The salt of the Earth," I call them here.

I watch them yet, in silent prayer,
Fond memories live – they still fly there,
For now I watch my grandchild play,
And thanks to them, I live today.

My watch must end, for the years grow late.
And I, like them, do calmly wait
To join that gallant band in Heaven,
And serve once more with Number Seven.

Sergeant Ignatius J. McCann, a Canadian Sergeant dying of cancer, sent the above poem to AVM Don Bennett, but it arrived too late; Don had died two days earlier.

'. . . He was a man who would not accept second best. Success on one day was only a springboard for greater success the next day. He was restless, imaginative and receptive to new ideas from his subordinates as I well know from personal experience as one of his squadron commanders. All who served under him knew he would never ask anyone to do anything he would not have been prepared to do himself – the hallmark of a true leader, and he developed a relationship with his squadrons and staff which outsiders rarely understood.'

Sir Ivor Broom, KCB, CBE, DSO, DFC, AFC, on AVM D.C.T. Bennett, CB, CBE, DSO, FRAes in The Marker, *Spring 1987.*

LANCS

Lancasters

Where are the bombers, the Lancs on the runways,
Snub-nosed and roaring and black-faced and dour,
Full up with aircrew and window and ammo
And dirty great cookies to drop on the Ruhr?

Where are the pilots, the navs and air-gunners,
WOP's and bomb-aimers and flight engineers,
Lads who were bank clerks and milkmen and teachers,
Carpenters, lawyers, and grocers and peers?

Geordies and Cockneys and Wiltshire moon-rakers,
Little dark men from the valleys of Wales,
Manxmen, Devonians, Midlanders, Scouses,
Jocks from the Highlands and Tykes from the Dales?

Where are the Aussies, the sports and the cobbers,
Talking of cricket and sheilas and grog,
Flying their Lancs over Hamburg and Stettin
And back to the Lincolnshire wintertime bog?

Where are the flyers from Canada's prairies,
From cities and forests, determined to win,
Thumbing their noses at Goering's Luftwaffe
And busily dropping their bombs on Berlin?

'Where are the bombers, the Lancs on the runways / Snub-nosed and roaring and black-faced and dour / Full up with aircrew and window and ammo / And dirty great cookies to drop on the Ruhr?' (via Tom Cushing)

Where are the Poles with their gaiety and sadness,
All with the most unpronounceable names,
Silently, ruthlessly flying in vengeance,
Remembering their homes and their country in flames?

Where are the Kiwis who left all the sunshine
For bleak windy airfields and fenland and dyke,
Playing wild Mess games like high cockalorum,
And knocking the Hell out of Hitler's Third Reich?

Where are they now, those young men of all nations,
Who flew though they knew not what might lie ahead,
And those who returned with their mission accomplished
And next night would beat up the Saracen's Head?

The Lancs are no more, they are part of legend,
But memory stays bright in the hearts of the men
Who loved them and flew them through flak and through hellfire
And managed to land them in England again.

The men who were lucky to live to see victory,
The men who went home to their jobs and their wives,
The men who can tell their grandchildren with pride
Of the bomber which helped to save millions of lives.

Audrey Grealy

'The Lancs are no more, they are part of legend / But memory stays bright in the hearts of the men / Who loved them and flew them through flak and through hellfire / And managed to land them in England again. The men who were lucky to live to see victory, / The men who went home to their jobs and their wives / The men who can tell their grandchildren with pride / Of the bomber which helped to save millions of lives.' (BAe)

'Flying their Lancs over Hamburg and Stettin / And back to the Lincolnshire winter-time bog . . .' (Sotheby's)

DOMINIONS

'Where are the Aussies, the sports and the cobbers . . . Where are the flyers from Canada's prairies . . .'

'The operational strength of the RAF reached a peak early in 1945, when a grand total of 505 squadrons were in existence. The influx of large numbers of men from the Commonwealth and from occupied Europe made a major contribution to the strength of the RAF, and almost a quarter of all squadrons were from overseas. Canada provided the largest contingent by far, with 45 squadrons numbered in the RAF sequence, followed by Australia (17), Poland (15), France (12), New Zealand (6), Norway (5), Czechoslovakia (4), the Netherlands (3), the USA (3), Belgium (2), Greece (2), and Yugoslavia (2). In 1945 these squadrons had a total of 8,752 operational aircraft, while the second-line and training units operated a further 18,691 aircraft. A total of 96 different types of

'Where are the Kiwis who left all the sunshine / For bleak windy airfields and fenland and dyke . . .' New Zealand High Commissioner Mr Jordan shakes hands with airmen of 75 (RNZAF) Squadron at RAF Mildenhall on 8 October 1942. (Dr Colin Dring)

aircraft were used by the RAF during the Second World War, many of them still being used in 1945. This total includes 36 American types, such as the Mustang, Harvard, and Dakota.'
Chris Hobson, Senior Librarian, RAF Staff College.

'Although there seemed to be no formal written policy, it became evident we RAAF aircrew serving with the RAF could expect two tours, at least one operational. I know of no-one who did two non-op tours, but some, eg., certain pathfinders, did two op tours. After repat' we could expect a second op tour against the Japs north of Australia with the RAAF. Of the five crews on my OTU course, we "colonials" got India, and my English navigator of course got India with me.'
M. Howland, RAAF Mosquito pilot.

'Canadian airmen operated on a two-tour system, and in intruder work the first tour would consist of 35 sorties, followed by a period usually as instructors at OTUs, or in operational planning at group HQ, then a second tour of 25 trips would round-out our operational obligation and we would be eligible for a posting home. Paul [Beaudet] and I did our first tour in 11 weeks and were still raring to go, so we applied for and were granted an extension to our first tour of 15 trips. We thought that by being very quiet about it we would be able to sneak in an extra ten sorties without anyone noticing and, therefore, we could say that we'd had done our two tours and so we'd get back home sooner. This plan didn't work, and as soon as we finished the extra 15 we were off operations.
George Stewart RCAF, 23 Squadron Mosquito pilot, and his navigator, Paul Beaudet, completed 50 trips; 45 to enemy aerodromes, 36 of them in Germany.

WARSAW CONCERTO

'Where are the Poles with their gaiety and sadness, All with the most unpronounceable names . . .'

In August 1944 the Russian armies were at the

approaches of Warsaw. The Polish Home Army under General Bor was persuaded to rise against the German occupation troops, but the Russians gave no support to the rising. Bor pleaded for all possible air support, but the Russians procrastinated until, on 8 and 9 August, seven Liberators of a Polish Special Duty Flight in Italy were allowed to make the almost suicidal 1,750-mile round trip over enemy-held territory. Then, 31 (SAAF) and 178 Liberator squadrons, 205 Group, were diverted from the invasion of southern France to fly "Warsaw Concerto" operations.

On 13 August Liberators of 31 and 34 (SAAF) and 178 Sqdns set out for Warsaw. Three were shot down, 11 failed to drop, but 14 dropped successfully. The next night 12 aircraft dropped safely but 8 failed to return. Of the 186 Liberators despatched on Warsaw Concerto operations, 92 succeeded in dropping supplies, 63 were unable to drop, and 31 failed to return. In all, the South African squadrons sent 41 aircraft, of which 11 were lost. Glorious failure though their mission may have been from a military standpoint, the brave and daring crews nonetheless brought everlasting credit to the name of the air force of which they were a part.

'Where are the Poles with their gaiety and sadness / All with the most unpronounceable names / Silently, ruthlessly flying in vengeance / Remembering their homes and their country in flames?' Flt Lt Marian Chelmecki, right, two confirmed victories in November 1940 and a Fw 190 in January 1945. (Sotheby's)

'Our first two operations were supply drops to Tito's Partisans in Yugoslavia. These were completed under codenames, flying about 100 ft over the DZs [dropping zones]. We dropped sugar, boots, rifles, and other supplies, and we could quite clearly see horses and carts coming to pick them up. During briefing for a raid on Yugoslavia we were told that there were three main Partisan groups involved – Tito, Mihailovich's men, and the Chetniks. We were told to watch out for the Chetniks because they were known to help the Germans look for downed airmen . . .

'The rear turret was fitted out for electrical flying suits, although we didn't get these until the end of the war. Owing to dampness in the tents we got a lot of shorting out. On one raid my gloves caught fire and I had to throw them out of the turret. After that I used only silk

On 13 August Liberators of 31 and 34 (SAAF) Squadrons and 178 (RAF) set out for Warsaw. (Dave Becker via SAAF)

gloves. My main clothing was two pairs of silk underpants, two vests, probably a shirt and RAF pullover, plus an inner suit. There was no way I could wear an Irvin jacket in the cramped confines of the turret.'

Frank Mortimor, Liberator tail gunner, 34 (SAAF) Squadron, 205 Gp, Italy, 1945.

'Flying Spitfires was the thrill of my life – there was nothing like it. However, the Mosquito came a close second and I was lucky to be flying what I consider to be the top two 'planes of the war. At Benson in 541 Squadron, in the Photographic Reconnaissance Wing, I completed 44 sorties on specially modified unarmed Spitfires. Later I

'At Benson in 541 Squadron, in the Photographic Reconnaissance Wing, I completed 44 sorties on specially modified, unarmed Spitfires . . .' (City of Norwich Aviation Museum)

joined 544 Squadron flying PR Mosquitoes and did 26 additional sorties for a total of 70 operational flights. PR was a very interesting job. We knew in advance of many occurrences. ie., the Dams raids, V2 rockets, etc. It was also one of the few jobs where one had the opportunity for independent action. We operated singly, and although we were briefed for definite targets, how and when we got there was largely up to us. We also had authority to divert to photograph any convoys, or other unusual targets spotted. We covered the whole of Europe in daylight from Norway to Gibraltar, and inland as far as Danzig and Vienna.'

Flight Lieutenant John R. Myles, RCAF.

". . . now I wonder if I packed my tooth-brush?"

(Handle with Care, *by D. Westmacott and R. Anderson)*

DEATH MARCH

'It was just after midnight on 6 February 1945 that we were awakened somewhat noisily by our own camp leaders with the news that we had to be ready to leave camp by midday and on foot. We were soon wide awake and making sure that a hot brew was being prepared . . . Our wanderings seemed to have little purpose and we were not progressing in any sort of a straight line. I got the impression that we were being moved around, still at the rate of about 15 miles a day, to the next convenient place to stop overnight. We had seen a fair amount of aerial activity, and on several occasions we had to dive into roadside ditches to avoid the attention of our own strafing fighters, mostly Typhoons. Before the pilots became aware of our existence we did sustain several fatalities and other lesser casualties, but it was soon realized who and where we were and thereafter the aircraft often flew alongside the column at low level, the pilots giving us a wave. The large formations of USAAF bombers were awesome, and we saw some of them shot down by defending fighters and cheered when the white canopies emerged to float downwards. One sunny day we had only just passed through a smallish town and had reached the outskirts when the Yanks came over and flattened the area we had traversed some ten minutes earlier.

'Quite unexpectedly one day we were told we were to be transported by train to a camp and, true enough, after about two miles we arrived at the railway station of Ebsdorf, where the usual cattletrucks awaited us. Other columns were

"'Ere y'are, Miss . . . you need it more than I do!"

(Handle with Care, *by D. Westmacott and R. Anderson*)

"Don't bother to unpack, we're moving tomorrow."

converging upon this railhead. We were packed into the trucks at least 80 men to a truck for the 50-mile journey, and consequently it was a case of standing room only. The doors remained closed throughout the night. It is not difficult to appreciate what a disgusting mess we were in when we were finally let out after about 20 hrs, for there were no toilet facilities and many of us were suffering from diarrhoea. I know of at least one Kriegie who did not emerge alive.'
Basil Craske, PoW, on the march from Gross Tychow in February-March 1945.

'Most PoW camps were put on a "death march" by March 1945 at the latest, although I think this a bit overplayed. It was rough, but I doubt if it compares with Bataan or that rotten railway in Burma.'
Geoff Parnell, PoW, 1945.

THE 10,000 PLAN

'To: Air Officer Commanding in Chief,
Headquarters, Bomber Command.
Huntingdon.
PFF/S.27/2/2/Air 8th January 1945

'Sir,
'I have the honour to submit the following suggestion.

It would appear that there is the danger at the present stage of the war of reaching a state of stalemate which might permit the enemy to recover his composure sufficiently to induce him to continue his hopeless but fanatical efforts indefinitely. It seems, therefore, that some impressive indication of our strength should be given to him. This might very well be combined with a declaration of intention to destroy the whole of Germany systematically up to the moment when that country is prepared to surrender unconditionally. Undoubtedly, the most important city remaining to Germany is its capital Berlin. Although damaged it still contains many vital industries and most of the Government departments. These latter in particular are only slightly damaged. The task, however, of destroying this target is a big one. In the ordinary sense of events it would requite a series of heavy raids which would suffer, after the first one, an expensive loss rate. Moreover, from past experience it would seem that crew morale would not maintain itself on a high level over a long period.

'The importance of striking a really devastating blow at the enemy would in itself be an incentive to crews to exert themselves to their utmost to achieve the best possible results. On a target as distant as Berlin such a spirit is essential.

'It is suggested, therefore, that while the nights are still long enough Bomber Command should make an all-out attack on Berlin. It is estimated that in one night this Command could drop 10,000 tons on the capital. This would require two trips for many of our aircraft and crews. This admittedly would be a great strain but is one well within the ability of the majority of our crews. Many German crews have been asked to fly two sorties per night against Great Britain, and although it is realized that these were shorter trips it is pointed out that the importance of this action would warrant the additional call on our crews . . . The losses involved are estimated at a probable 4 per cent on the first wave and 2 per cent or less on the second wave. It is realised that as a total number of aircraft this sounds an appreciable loss figure, but it is pointed out that the destruction wrought in the German capital would provide an object lesson at least comparable with the Battle of Hamburg which had such a devastating effect on the morale of the German people at that time.

'At this stage of the war something out of the ordinary is urgently required. I therefore submit this suggestion for your favourable consideration.'

(Signed) Don Bennett
Air Vice Marshal, Command
Pathfinder Force (No.8 Group)

'On 14 February we witnessed the bombing of Dresden. We were supposed to sleep in a barn on the outskirts, but the sirens went. The RAF came in and bombed all night, the Yanks came in and bombed all the next day, the RAF came in and bombed all the second night. The German guards told us that more than 100,000 civilians had been evacuated from the Russian front at Breslau. They never did get an accurate count of their dead from that raid on Dresden.'

Boston Patterson, RCAF, PoW on the march in 1945.

'Never before has an Allied air raid produced a reaction such as we see now. The familiar epithets – "terrorflieger," "luftgangster", and "kindermörder" – are spread lavishly across the front pages of the German dailies in condemnation of the Allied bomber crews.

'Oddly enough, there is no great hostile reaction towards us on the part of the German guards, many of whom lost families in the Dresden "terrorangriff".

'Most of them, particularly those who experienced the holocaust, seem too shocked and dazed to be capable of normal emotions like hatred and a desire for revenge.

'One guard who was on duty in the stalag when Dresden was bombed applied for leave to see if his family were still alive. On arrival in Dresden he not only failed to locate his family or his house; it was also impossible to identify the street and the suburb. When he left Dresden, eastbound German troops stranded in the burning city were turning their flamethrowers on the ruins to cremate the thousands of trapped bodies and prevent outbreaks of disease.'

Geoff Taylor, RAAF, PoW, Stalag IVb, Muhlberg-on-Elbe, Saxony, 25 miles from Dresden. Piece of Cake *(George Mann, 1956).*

(On 13/14 February 1945 Dresden was bombed in two assaults, the first by 244 Lancasters, the second by 529. Some 2,659 tons of bombs and incendiaries fell on the town. On the next day, 311 bombers of the 8th Air Force followed with a raid. Over 32,000 people were killed).

'The destruction of Dresden remains a serious query against the conduct of Allied bombing.'
Winston Churchill.

'The war was now in Germany, and we were in an area of constant alert. There would be no warning siren. When it sounded, it meant bombers overhead. One day, a flight of six Mustangs spotted us and mistook us for a column of German reinforcements. They came in low, with rockets and guns blazing, and wiped out more than 150 of us, not counting the wounded. On 15 March we arrived at Ziegenheim, about 20 miles from the Rhine. The deep rumble of war was now to the west of us. Our original column of 3,000 men was down to 1,000. Most of us were desperately ill with dysentery and starvation. When we stood up we moved very slowly, or we would fall. When the pangs of starvation set in, you first feel the straight gnawing of hunger; then the craving for something sweet; then the craving for fat; then the craving for salt, and this is when you get dizzy.

'We slept under canvas, and didn't march again. We couldn't have marched, even at gunpoint. The Yanks surrounded our position and recaptured us on 31 March – Good Friday. The RAF flew us to England early in April. We had not shaved, had a bath, or even had our boots off in more than ten weeks. Those of us

who could spare it lost about 60 lb. Those who couldn't spare it – just didn't.'
Boston Patterson, RCAF, PoW on the march in 1945.

VARSITY

Hymn to Airmen
Inspire, O Lord, our men who fly
Their winged chariots on high,
Across the dark and tortured sky.
Take them whereso'er they fare
From all the dangers of the air.

Group Captain E.B.C. Betts, July 1940.

'In the Horsa glider, the two pilots sat side-by-side in a cockpit not unlike a conservatory, with almost 360° view. There were two control columns of the spade-grip type, and the only instruments were an altimeter, airpseed indicator, a compass, and another colloquially called the "angle of the dangle". It was a Heath Robinson device in which thin cord only a few feet long was connected to the tow rope, and this indicated, in an illuminated panel, where the glider was situated in relation to the tug aircraft. This was essential when flying in cloud or at night. The Horsa could carry 28 fully armed troops, and the idea was to land them at some strategic point, such as a bridge or enemy gun emplacement, possibly at night.

'In the summer of 1944 over 200 graduates from 4 BFTS [British Flying Training School], Falcon Field, helped the Army as glider pilots. The newly trained RAF glider pilots were split into two groups. Roughly half of them were allocated to UK airfields for the Rhine crossing, and were attached to No. 1 and 2 Wings of the Glider Pilot Regiment; making up the crews with survivors from Arnhem. The remainder were allocated to man the six squadrons of RAF Glider Pilots, 668–673, which made up 343 and 344 Wings, founded in India for airborne assault in South-east Asia. Each of these squadrons had an establishment of 80 Hadrian gliders – known to the Americans as Wacos [the name of the manufacturer].

'Back in the UK, the time was drawing near for the Rhine Crossing, and between 20–22 March 1945, RAF Bomber Command and the US 8th and 9th Air Forces made 16,000 sorties over the area and dropped 49,500 tons of bombs. At 17:00 hrs on 23 March the entire artillery of the British 2nd Army opened fire on enemy positions and maintained the barrage of shells until 09:45

'The Horsa could carry 28 fully armed troops, and the idea was to land them at some strategic point, such as a bridge or enemy gun emplacement, possibly at night.' (Boulton Paul)

on 24 March. The total airborne force consisted of 1,795 paratroop carriers and 1,350 towing gliders, escorted by 889 fighters; a total of 3,989 aircraft.

'With this massive formation, the Allies were able to land all 21,680 glidermen and paratroopers of the British 6th Airborne Division and the American 17th Airborne Division in 2 hrs 36 min. The Varsity drop was to be the largest single airborne attack made by either side during the war – even larger than the one made in Holland.'

Herbert Buckle, glider pilot and 4 BFTS cadet, writing in The Falcon, *the 4 BFTS newsletter.*

'We crossed the Rhine enveloped by the now-famous "Monty's smoke-screen". We couldn't see the river but we overflew Hamminkeln and I could see the River Issel and our bridge. By this time the AA fire was heavy. Tracer bullets crept up towards us and then flashed past, and the sky was filled with exploding shells. I saw the leading glider, flown by Capt Carr, break up in the air and the men falling out. Other gliders in the North Bridge Group were hit. Then it was our turn.

'Our tug took us right over the Dropping Zone and they shouted "Good Luck" over the intercom as we cast off. The glider with our jeep on our starboard disappeared, and then a shell exploded under us. I pinpointed our spot in a field as per the photograph we had been given, and Pete Ince dived for it.

'We left the flaps until the last moment, but when I pulled the flap lever nothing happened – our air supply had been cut. We gained speed and the airspeed indicator went off the clock! It took both of us all our strength to pull out of that dive. We hit the ground hard, raced along at practically zero feet, and I remember the German machine-gunners ducking for cover as we almost decapitated them. Our excess speed made us climb away and, as we tried to turn back into our field, we hit the top of a belt of trees. These helped to save us, and as the port wingtip touched the ground we came to a sudden stop and I was thrown out through the nose and made a dent in the ground.

'Those of us who survived were pinned own under the wreckage by machine-gun fire. Soon the glider began to burn, and as the flames drew near our cargo of explosives we had little option but to race across the field to a trench we could see in the distance. It was only then that I realised the two soldiers lying on either side of me had both been killed since we landed. We all

'We crossed the Rhine enveloped by the now famous "Monty's smoke-screen".' (via Frank Crosby)

made the trench without further casualties, but discovered that the said trench was an open sewer. We sank in up to our waists. For once in our lives we were glad to land in the . . !

'Jock Davies and I looked for arms and picked up a Piat gun. We were then told to fire it toward a gun emplacement. After a large group of the enemy, consisting of German Home Guard and an SS officer, had surrendered, I found myself with one paratrooper in charge of them. As we marched them off I could not resist ordering them to give an "eyes right" to an Airborne colonel, who, seeing my RAF pilot's brevet, was kind enough to congratulate the RAF pilots on their showing.

'Hamminkeeln was still receiving the occasional shell as we began to pull out. When we drew near the pontoon bridge across the Rhine, the Commandos who had crossed the river in boats lined up to cheer us on our way out. We were taken to Helmond, where the Dutch people looked after us very well. Then to Eindhoven and flown back to RAF Down Ampney on 30 March. There we were met by HM Customs, who wanted to know where we had spent the past few nights! The answers they received were varied to say the least. Tempers were short, and they soon diplomatically ran out of forms. Maybe the threat to burn their hut down helped.'

Sergeant Pilot E.W. Ayliffe, writing in The Falcon, *the No 4 BFTS newsletter.*

The British glider force consisted of 440 Airspeed Horsas and General Aircraft Hamilcars. Anti-aircraft fire was murderous, and only 80 British gliders landed unscathed. The assault resulted in an outstanding victory. It was not however, achieved without cost, for the Glider Pilot Regiment had suffered 101 pilots killed, of whom 61 were RAF glider pilots.

HOME BY CHRISTMAS – NEXT YEAR?

'We'll be home for Christmas, if only in a dream.'

'In April 1945 the bulk of RAF NCO prisoners of war were held in Stalag 357 at Falingbostel, lying between Hamburg and Hanover in north-west Germany. The camp also lay under one of the main routes flown by the RAF at night, and by day the bombers and fighters of the US 8th Air Force.

'During the intensive air operations in the closing months of the war, the prisoners saw many aircraft from both sides fall to the ground, and sometimes a single or a cluster of parachutes hanging apparantly motionless in the air.

'During night raids many relived their own experiences as they watched searchlights form into cones, encircling a single bomber now a target for the guns. Over the drone of the bomber stream muttered oaths were heard, and cries of: "For Christ's sake, weave" as they saw stabbing points of light from bursting flak shells within the cones. All too often a huge conflagration followed the firework display, signifying the end of another bomber. One wondered – and hoped – for the crews.

'We had been at Falingbostel since August 1944, having been evacuated from Hydekrug, East Prussia, in June, and Torun in Poland two months later. The guns of the advancing Russian armies could be heard at both the latter camps, and now Allied guns were moving close to Falingbostel from the west.

'It had been a difficult winter for the prisoners. The Red Cross lifeline was broken, and everyone now existed on the bottom ration scale, designed to maintain little more than subsistence and reserved by the Germans for non-productives and criminals. Fuel was almost non-existent, and the large huts packed with three-tiered bunks were constantly wet with condensation. Even sleeping, lying fully dressed in our bunks, was made uncomfortable. As a reprisal for German prisoners allegedly being forced to sleep on sand in the Middle East, our straw palliases had been removed. To make matters worse, most of our bedboards had already been taken and burned for fuel. Many times during the night startled oaths told where someone had fallen through gaps in their bedboards on to unlucky neighbours beneath.

'Although morale remained high, the main contributory factor to this being the BBC news received on hidden radios, there was a growing impatience for the war's end, particularly among the prisoners of four and five years' standing. It seemed as though the nearer the end came in

"I want a suitable present for my son . . ."

sight, the more difficult it became to face the intervening period and its many uncertainties.

'As the sound of guns grew louder, rumours abounded about being marched off into the interior, possibly as hostages. Many of us had in mind the possibility of a last-minute revenge by the Germans against airmen responsible for the destruction of German cities. On 6 April 1945 our camp leader, Dixie Deans, was informed by the German commandant that we were to march out of the camp. Dixie, who had a splendid instinct for right decisions, advised compliance and go-slow tactics. The columns formed up very slowly, and it was well into the afternoon before the group in which I was a member moved out, each man carrying his worldly possessions. We dawdled along the road; our guards, who appeared to have little enthusiasm for their task, were nevertheless treated with caution, as their attitude and temper seemed uncertain.

(Handle with Care, *by D. Westmacott and R. Anderson*)

'It was dark when we reached our first night stop, a large barn only a mile or two from the camp. On this night there was a lot of air activity, and soon a flare was dropped, lighting up the countryside as if in daylight. An aircraft was then heard diving down, and it passed very low over our barn. A few seconds later there was a flash of light, followed by a loud explosion. The aircraft, believed to be a Mosquito, had hit some trees and crashed. Next morning a German brought along the identification tags of the two dead airmen.

'The next few days were a kaleidoscope of apparently meaningless wanderings around the German countryside with the noise of war a constant background; the distant rumbling of guns and Allied aircraft flying overhead. The latter were a particular worry to us, particularly the roving Spitfires looking for ground targets. We would see them wheeling in the distance, the sound of cannon reaching us after the aircraft had again attained level flight.

'We would wave frantically as the Spitfires banked over and circled above us. Much to our relief they always flew away, our ragged appearance probably satisfying the pilots that we were not a German column. Not so fortunate, though, for one group ahead of us. They were caught by Typhoons as they sat around eating some recently supplied rations. Thirty-seven died, including one a prisoner since April 1940. It was hard to affect the usual apparent indifferent shrug – so near the end and by our own Service as well.

'Most nights were spent in fields, the weather fortunately remaining mild and dry. One night, as the bomber stream flew towards us, a brilliant "Christmas tree" marker of red and green was dropped right above the field. The aircraft wheeled to the north, and some time later flashes of light and rumblings of sound showed where the target lay.

'During our wanderings we criss-crossed long columns of unescorted French prisoners and displaced persons pushing and pulling a weird assortment of prams, trolleys, and carts. We envied them their apparent wealth of food and freedom, our meagre rations being supplemented by forages against potato clamps, beet stores, and meal bags. Once I returned in triumph after foraging in a farm store, but alas the oats I brought back were heavily laced with fishmeal and inedible.

'One morning we found ourselves near a fast-flowing steam. We stripped off and got into the chilly water, the first real wash since we left the camp a week before. On one occasion we came across the remains of a German column which had been attacked from the air. Several dead horses lay alongside the road. We fell on these beasts with our knives and later that day enjoyed horse steaks grilled over fires which surprisingly enough the guards allowed us to light.

'As was customary in PoW camps, four of us pooled our resources into a combine, and in discussing our situation had decided that we should watch out for the right moment to escape from the column. We bore in mind the German warning that those dropping out would be shot, as well as the possible dangers in meeting members of the German civilian population, many of whom had been evacuated from bombed cities. On the other hand, there was no guarantee that the Germans would release us safely at the end. One of the combine began suffering the effects of privation and, in spite of our pleas, could go on no longer. In answer to our warning about being shot, he merely said: "Sod it, I don't give a damn," and sat down by the roadside. The German guard spoke to him, and much to our relief moved on, as did the next guard. He was lucky to be eventually picked up by some Germans, taken to Luneburg and lodged in the jail, where he spent the next four days before being released. That night, 14 April, was spent in a barn. The guards advised us to get a good rest as it was their intention next day to cross the River Elbe, some 30 km distant.

'The combine agreed that this was the moment to get away. We felt that once across the Elbe it would not be possible to get back again across the river. That night we burrowed deep

(Handle with Care, *by D. Westmacott and R. Anderson*)

"Aren't men beasts!"

into the straw, and next morning waited until everyone had gone before emerging. We were sitting just inside the barn, brushing ourselves down and gathering our things together, when a German soldier armed with a rifle walked through the doorway. This was the sort of encounter we feared most, but fortunately he merely asked what we were doing there. We said that we were sick and had to rest. He appeared to be satisfied with our answer, and after a few further exchanges went on his way, much to our relief.

'We started off, keeping to paths and by-roads. Except for aircraft wheeling overhead, the countryside seemed strangely empty. We reckoned the Allied forces were only a few miles away and hoped to avoid any retreating German troops. After some time we came across a horse-drawn cart containing some sick British prisoners and others walking alongside. They were making their way toward a central point for sick prisoners at Melbeck, some 10 km away. Melbeck, a small village 12 km south of Luneberg, lay in the general direction we were taking. We joined the party and eventually arrived later that day. In a farmyard at the end of the village a busy Army doctor was looking after a number of men lying on the floor of a barn. Occasionally a nervous looking farmer's wife brought over milk, her three children peeping around the door. We were glad of the opportunity to rest, and planned to stay a couple of days before pushing on.

'In the farmyard under a tree was a group of young German soldiers, the crew of a mobile flak gun. They were friendly enough, and showed us the damage to their vehicle after an air attack. The driver was particularly pleased that he had escaped injury in spite of severe damage to his compartment. The crew were now sitting the war out and waiting for the end. That night they drove off with a cheery 'auf Wiedersehen!'. In a stable a horse with a blown stomach through eating grain was dying. Someone borrowed a gun and put the beast out of its misery. From then on there was plenty of meat stew available at all times.

'Nothing moved on the road outside, except on one occasion a fire engine went past at high speed towards Luneberg. It had travelled only a short distance when it was caught by three Spitfires firing their cannon when over the farm. The cannons' roar reverberated among the farm buildings. There was a flash of fire and smoke as the fire engine crashed into the side of the road. Nothing else moved. It was obvious that the end was now near. The sky that night of 17 April glowed from fires quite near. It was wise to stay where we were.

'Early in the morning of Sunday 18 April the sound of approaching tanks was heard. White flags were already hanging from windows in the village. We rushed to the road and saw a mass of tanks and trucks piled with troops. Soon we were part of a milling mob with men of the King's Shropshire Light Infantry [KSLI]. We climbed on to the tanks in a delirium of excitement, our pockets being filled with cigarettes, chocolate, and rations.

'A member of the KSLI invited me to go with him checking houses for hiding Germans. We marched into a house but I did not like the sight of these cowering, frightened people, who were now completely at the mercy of their conquerors, even though I speculated that these same people had probably no such qualms when their own troops were marching in triumph across occupied countries.

'The next few days are a misty memory of shattered buildings where stands had been made, empty shell cases where tanks had stood and fired, crashed aircraft, and the litter and waste of war everywhere. And everywhere, white flags. We cadged lifts in returning supply trucks, spent a day or two at Nienburg and then on to Rheine, the airfield a scene of utter destruction following Allied air raids. But craters had been filled in and RAF Dakota aircraft were busy flying in and out. Prisoners were now arriving in large numbers and waiting for flights to take them home. I just missed the last aircraft leaving that day for England, and was standing around looking at wrecked German aircraft in the shattered hangars, and at a Dakota which had run into a crater and tipped over, when a Canadian pilot asked me how long I had been a prisoner. On learning five years, he said: "Heck, I guess you'd like to get back then. If you don't mind riding on crates, I'm just leaving for Wing, one of the PoW reception centres." Those of us in the vicinity accepted the offer with alacrity.

'I wondered how I would feel about flying again. It was three weeks short of five years since I was last in an aircraft, and from that occasion I still had the feel of cordite, spilling oil, and glycol, smoke, and fire in my nostrils. We lifted into the air and the intervening years disappeared. The drill came back, wheels up, flaps up, change from fine pitch. I envisaged the

HOME AT LAST

". . . funniest thing I ever heard!! . . ."

(Handle with Care, *by D. Westmacott and R. Anderson*)

wireless operator sending his messages, and wondered whether the procedures had changed very much. I lay back on my crate and looked at the roof, and wondered whether the last five years had happened. My dirty khaki battledress was enough to remind me it had. I looked down on the fields of England and soon we were landing at Wing. I called to the pilot: "Thanks for the ride". He laughed. "Well don't stay away so long next time." I jumped from the aircraft. I was on English soil again. It was 23 April 1945, just five years and two months, a fifth of my life, since I last left England to return from leave to my squadron in France. I would be home for Christmas this year, the first for seven years.'
Len R. Clarke, PoW, shot down 14 May 1940.

'I had heard from captured 101st "Screaming Eagle" paratroopers about the German

(Punch)

"They met exactly two years ago on a float in the Channel."

breakthrough in the Ardennes and Bodenplatte. They were totally demoralized, and believed the Germans could go through all the way. The Russians, meanwhile, were coming like the Scotch Express. So, in March 1945, during the march westwards away from the advancing Russians, Nick Green, a prisoner since being shot down in a Whitley in 1941, and I, made our getaway from the column. We would hide and wait for the advancing Russian Army. We hid for three days and nights in bitterly cold conditions in swede clamps, 30 ft long by 10 ft wide, pulling straw over our bodies for warmth. It was no good. We both got badly frostbitten limbs and gangrene set in. On the third day I gave myself up. Nick was in very bad shape.

'We were taken to Neubrandenburg Kriegsgefangenen Lazarett (hospital) where a Polish naval surgeon, Capt Mickoski, who had been captured at Gydnia in 1939, operated on us. He was assisted by a Serb of the Royal Yugoslav Army. Communication was rather difficult, and the two of them argued nineteen to the dozen. Our operations were carried out with spinal anaesthetic, so we were conscious all through the operations. Nick and I had a double Chopart's [operation] to remove our toes and front part of our feet. Nick was in worse condition, the gangrene having spread up his legs. His next operation was a double below the knee and then a double above the knee. After the third operation he died of shock. I had been down for only 20 months and was 23 years old, so I could perhaps better stand the amputations than Nick, who was in bad shape, having been down since 1941.

'When the Russians overran the hospital I received treatment from one of their nurses. She had been fighting for three years and carried a sniper's rifle. Apart from her nursing duties she would direct traffic with biff-bats when "off duty"! I spent three weeks with the Russians. The nurse gave me blood transfusions Russian style with a syringe to pump blood into my arm! It was the same needle used by everyone else in the ward!

'Back in England I was sent to EMS at King's College Hospital in Epsom, where they sorted out my gangrene. I was nursed by Norah Hudson of the Physiotherapy Service, who was attached to the RAF. It was against the rules, but she and I sneaked out through a hole in the

'Prisoners were now arriving in large numbers and waiting for flights to take them home.' (Imperial War Museum)

hedge at night, Norah pushing me on her Hercules bicycle to the Marquis of Granby pub. Later, we cycled around Chessington to help strengthen my legs. We agreed to marry three weeks after meeting, but because of Service delays we married in 1946.'
Len Bradfield, 49 Squadron.

LIBERATION

The tumult and the shouting dies;
The captains and the kings depart;
Still stands Thine ancient sacrifice,
A humble and a contrite heart.
Lord God of Hosts, be with us yet,
Lest we forget – lest we forget . . !
Kipling's *Recessional*

'We saw the Spitfires more in the last years of the war. They were really very daring. If you saw them coming down it was so very exciting. The Dutch of course were very much in favour of the Tommies, and were so happy that they got rid of the Germans for us. I have never seen anything more exciting. They would come DOWN, and then they were UP again! It was fun, even if we were feeling miserable about the whole trouble. It gave us a lift. They were our liberators. We waved, but they had no time to look at us. It was little things like this that made it bearable to be occupied. It was fun to do this. It scared the Germans. We were just so thrilled they were coming. We would never have known when the Germans might leave. If it wasn't for them they may have stayed another five or ten years. Liberation was so fantastic. Free again!"
Twenty-six-year-old Christina van den Born, Holland 1945. (In late April 1945 the RAF and USAAF began Operation Manna to air-drop food and medical supplies to the starving Dutch population. RAF DZs were marked by aircraft of the PFF.)

'The pub was full, there was considerable excitement, and the beer was being consumed in great quantities. As the evening progressed it became obvious that the celebration of the ending of hostilities was to go on well into the night. Then there was a call – "Let's have a bonfire!" The crowded bar emptied, the drinkers taking the bar furniture out with them into the road. dad and I stood some few yards away and watched the furniture being piled up in the centre of the road. The next thing was that it was going up in flames, accompanied by a loud cheer from those gathered around it. We watched for a few minutes, then dad turned away and headed towards home. I followed and caught up with him. We walked in silence for a while, and then dad said: "I didn't like that". I told him that it was not what I wanted to see either . . ."
Sergeant Roland A. Hammersley DFM, 7 May (VE Day) 1945.

'Men and women of Bomber Command . . .

'To all of you I would say how proud I am to have served in Bomber Command for four and a half years, and to have been your Commander-in-Chief through more than three years of your saga.

Your task in the German war is now completed. Famously you have fought. Well you have deserved of your country and her Allies.
Special order of the day, 12 May 1945, from Air Chief Marshal Sir A.T. Harris KCB OBE AFC, Commanding-in-Chief, Bomber Command

AIR VICTORIA CROSS AWARDS, 1939–1945

'All of the soul
Of man is resolution which expires
Never from valiant men till their last breath."

'Our generation caught a packet and the Air Crew Europe medal fetches fourteen quid or so in the junk shops. Other campaign medals make 30 shillings apiece.'
Geoff Parnell, air gunner, 1977.

'Any operation that deserves the VC is in the nature of things unfit to be repeated at frequent intervals.'
ACM Sir Arthur T. Harris

The Allied Air Forces: A Dedication
How can we praise the brave young men
Who flexed their new and untried wings
And flew, like eaglets, free and bold,
No fettered slaves or underlings?

How can we tell of loyal men
Who worked long hours upon the ground
In drizzling cold or tropic heat,
To serve their crews in honour bound?

How can we laud the gentle girls
Who nursed them, fed them, mastered arts
Of radar, plotting, ground control
And tender care of engine parts?

How can we honour those who came,
Through hardship, danger, fear and pain,
From Europe's crushed and bleeding lands
To spread their wings and fight again?

How can we show our gratitude
To those who, of their own accord,
Came from our far-flung Commonwealth
To help us wield our shining sword?

How can we thank the men who crossed
The great Atlantic, deep and wide,
And left their New World far behind
To fight for freedom at our side?

A generation, selflessly,
Gave up the flower of its youth,
A sacrifice so quietly made
That we might live in peace and truth.

Recipient	*Squadron*	*Aircraft*	*Action*	*Award*
Garland, Fg Off Donald Edward, pilot.	12	Battle	12.5.40	11.6.40*
Gray, Sgt Thomas, observer.	12	Battle	12.5.40	11.6.40*
Learoyd, Act Flt Lt Roderick Alastair Brook, pilot.	49	Hampden	12.8.40	20.8.40
Nicolson, Flt Lt Eric James Brindley	249	Hurricane	16.8.40	
Hannah, Flt Sgt John, WOP-AG	83	Hampden	15/16.9.40	1.10.40
Campbell, Fg Off Kenneth, pilot	22	Beaufort	6.4.41	*
Edwards, Acting Wg Cdr Hughie Idwal DFC	105	Blenheim	4.7.41	22.7.41
Ward, Sgt James Allen RNZAF, 2nd pilot	75	Wellington	7.7.41	5.8.41
Nettleton, Act Sqn Ldr John Deering, pilot	44	Lancaster	17.4.42	28.4.42
Manser, Fg Off Leslie Thomas RAFVR, pilot	50	Manchester	30/31.5.42	20.10.42*
Middleton, Flt Sgt Rawdon Hume, RAAF, pilot	149	Stirling	28/29.11.42	15.1.43*
Malcolm, Act Wg Cdr Hugh, Gordon	18	Blenheim	4.12.42	27.4.43*
Newton, Flt Lt William Ellis RAAF			16.3.43	
Gibson, Acting Wg Cdr Guy Penrose DSO DFC, pilot	617	Lancaster	16/17.5.43	28.5.43
Trigg, Fg Off Lloyd Alan DFC RNZAF, pilot	200	Liberator	11.8.43	*
Aaron, Flt Sgt Arthur Louis DFM, pilot	218	Stirling	12/13.8.43	5.11.43*
Reid, Act Flt Lt William RAFVR, pilot	61	Lancaster	3/4.11.43	14.12.43
Barton, Pt Off Cyril Joe RAFVR, pilot	578	Halifax	30/31.5.44	27.6.44*
Hornell, Flt Lt David Ernest RCAF	162	Catalina	24.6.44	*
Cruickshank, Fg Off John Alexander, pilot	210	Catalina	17.7.44	
Cheshire, Wg Cdr Geoffrey Leonard DSO DFC RAFVR, pilot	617	Lancaster		8.9.44
Lord, Flt Lt David Samuel Anthony DFC, pilot	271	Dakota		19.9.44*
Thompson, Flt Sgt George RAFVR, W/Op	9	Lancaster	1.1.45	20.2.45*
Palmer, Act Sqn Ldr Robert Anthony Maurice DFC RAFVR, pilot	109	Lancaster	23.12.44	23.4.45*
Swales, Capt Edwin DFC SAAF, 'master bomber'	582	Lancaster	23/24.2.45	24.4.45*
Bazalgette, Act Sqn Ldr Ian Willoughby DFC RAFVR 'master bomber'	635	Lancaster	4.8.44	17.8.45*
Gray, Lt Robert Hampton DSC, RCNVR			9.8.45	
Jackson, Sgt (later W/O) Norman Cyril RAFVR, Flt Eng	106	Lancaster	26/27.4.44	26.10.45
Trent, Sqn Ldr Leonard Henry DFC RNAZF, pilot	487	Ventura	3.5.43	1.3.46
Scarf, Sqn Ldr Arthur Stewart King, pilot	62	Blenheim	9.12.41	21.6.46*
Mynarski, Pt Off Andrew Charles RCAF, mid-upper gunner.	419	Lancaster	12/13.6.44	11.10.46*

* Posthumous award

All tears were shed long years ago,
But memories last for all our days,
And so, with pride, we make our vow
To keep faith with them always.

Audrey Grealy
Written to mark the occasion of the unveiling of the monument to the RAF and Allied Air Forces on Plymouth Hoe, 3 September, 1989, the 50th anniversary of the outbreak of the Second World War.

'. . . They will hammer their swords into ploughshares, and their spears into pruning hooks. Nation will not lift up sword against nation, and never again will they learn war.'
Isaiah 2; 2–4.

THE STATELY 'DROMES OF ENGLAND

The Stately 'dromes of England are just a trifle bleak,
From Biggin Hill to Thurso, from Finningley to Speke.

The Stately 'Dromes of England; Flt Lt John Mark, 1943.

'Over 1,000 airfields were used by the RAF in the UK alone during the Second World War, so it is not without some justification that Britain was referred to as "a vast aircraft carrier anchored off the north-west coast of Europe". By the end of the war some 360,000 acres of land had been occupied by airfields and a staggering 160 million square yards of concrete and tarmac had been laid down.

'In 1945 the RAF's personnel strength consisted of a total 190,256 officers and 1,006,267 airmen . . . After nearly six years of war the RAF had developed out of all recognition from its pre-war existence. By 1945 the RAF had become the third largest air force in the world behind those of the USA and the Soviet union. The RAF had been involved in virtually every campaign throughout the war and had amassed a wealth of experience in a wide variety of roles and missions.

'Victory had not been achieved without a high price. A total of 70,253 RAF aircrew were lost on operations between 3 September 1939 and 14 August 1945, with no fewer than 47,293 being lost from Bomber Command alone. The Air Forces Memorial at Runnymede commemorates the names of 20,435 airmen of the RAF who were lost during the Second World War and who have no known grave.'
Chris Hobson, Senior Librarian, RAF Staff College, Bracknell, writing in 1995.

'In memory of Flight Sergeant Ken ——— ,
lost in action, April 1945.
We never forget – wife ——— and son Peter.
"Dad, the grand-daughter you never knew
gets married next week."'
Inscription on a card attached to a wreath, Runnymede, 1994.

'For all those here who failed to return . . .' (Author)

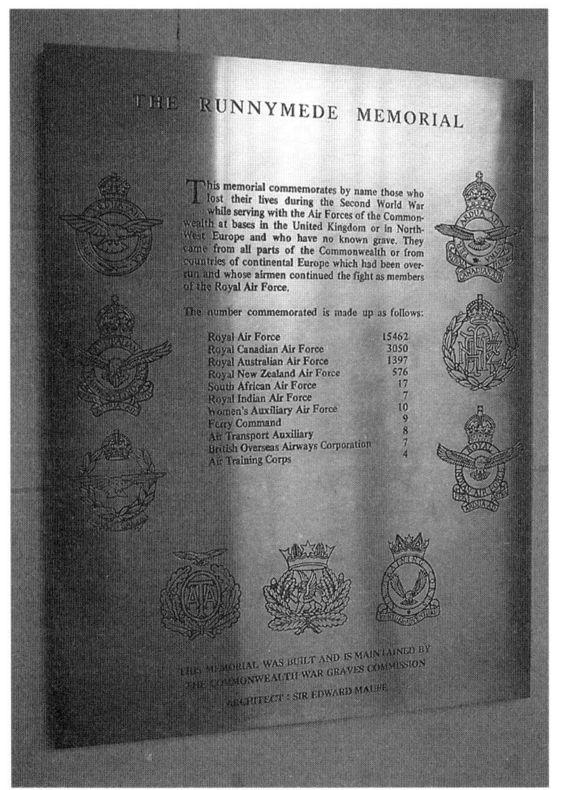

'The Air Forces Memorial at Runnymede commemorates the names of 20,435 airmen of the RAF who were lost . . . and who have no known grave.' (Author)

'For all those who failed to return – especially for the "Prosper" Network – and most of all for Noor Inayat Khan – inexplicably abandoned. Not forgotten – For Lilian Rolfe.' *Inscription on a card attached to a wreath, Runnymede, 1994.* (Noor Inayat Khan ['Madeleine'] and Lilian Rolfe were WAAFs, trained as special agents and dropped behind enemy lines. Both were executed at Dachau concentration camp.)

Solitary Toilet

Through what had been an airbase I chanced to drive one night when a white ceramic toilet came to view
The sky was almost cloudless and the moon was shining bright which made this aged receptacle seem new
Either new or ghostly but there was no fear, no chill, and no other indication spooks were near
And I know spectral bodies have not the need nor will for visiting of toilets to park the astral rear
But the toilet in its glory, alone, bereft of wall,

raised one foot above the ground, was standing like complete
Alas without a cistern, no pipe, no chain at all, and some damned thief had stole its wooden seat,
Then later, due to business, I often passed that way, which proved the solid toilet was on site
Midst little stunted bushes it was visible by day and in sunshine after rain reflected light.
Now I'm of curious nature and I often wondered why, how come one single toilet to be there?
When military build toilets the number rules apply, and sometimes then one cannot find a spare!
So was this single edifice for those of senior rank? Or was the Base Commander rather coy?
Did perfumed disinfectant ensure it never stank? Wasn't it for a lady within the Base employ?
But single bloody toilets are not usual on a base, at least they weren't as far as I'm aware;
This stupid problem puzzled as I drove by at pace, until some vandal smashed it and now it is not there.
It could have stood for centuries without this sinful act and then become a shrine one future year.
Historians could have mounted, with words of simple tact, a sign to read that General ????????? sat here!

Jasper Miles

Old Bomber Base Revisited; A Pilot's Pilgrimage to the past

Deserted, abandoned, an airfield spans the lonely heath,
Unkempt broken runways sprout their share of grass and weeds
Bare dispersal pans of circular concrete sit, now empty,
Lacking the black silhouettes of the bombers,
Which used to squat, etched against the darkening sky.

Empty pre-fab huts, with broken glassless windows,
Gaze sightlessly out at overgrown hedgerows,
And seem to echo back the voices and laughter of youths,
Who, in blue, once rode the skies to destruction and death.

The wind sighs in lonely desolation as if recalling
The vibrant roar of countless Merlins, coughing
Puffs of blue smoke to be whirled away
In the swirling propwash of many Lancs,
Ponderously thundering into the clouds and, when massing,
Made the very earth tremble with their passing.

'. . . the toilet . . . Alas without a cistern, no pipe, no chain at all, and some damned thief had stole its wooden seat . . .' (Caroline Frick)

Where groundcrew kept tally of every departure,
And muttered a prayer for each aerial charger;
Throughout the long nights their vigils maintained,
'Til in the grey dawn, their visages strained,
They counted the losses.

A strange way of life, to protect a way of living!
Sacrifice demanded, and the ultimate too often given.

A lonely figure walked to where a runway ended,
Thoughts deep in the past as his spirit blended.
Seeing this airfield as he once knew it.
Remembering well! He'd flown, and lived through it.
Exorcising ghosts, he roamed o'er the acres,
Recalling faces, nicknames, of givers and takers.

Pilots, navs and flight engineers too,
Wireless ops, gunners, from each motley crew.

But the visions all vanished. The noises all dimmed,
'Til all that remained was the sigh of the wind;
A creaking window, the rustle of grass.
Returned to the present. Bade Adieu to the past.

Jim McCorkle, ex-RAF pilot

'Deserted, abandoned, an airfield spans the lonely heath . . . now empty / Lacking the black silhouettes of the bombers / Which used to squat, etched against the darkening sky.' (Author)

'Empty pre-fab huts, with broken glassless windows / Gaze sightlessly out at overgrown hedgerows / And seem to echo back the voices and laughter of youths . . .' (Author)

Bomber Aircrew – Times Past

Life was a fleeting moment when
We lived from day to day,
A morning dawned, the sun broke through,
We savoured every ray,
For well we knew that, with the dusk,
There was a price to pay,
When we were young.

The dangers that we faced became
A common bond to share,
The friendships forged upon such fire
Were rich beyond compare,
So many of them all too short,
Their loss so hard to bear,
When we were young.

We lived our lives up to the hilt,
We laughed and loved and prayed,
We learned to crack the flippant joke
If we should feel afraid,
These things were all accepted
As by us the rules were made,
When we were young.

So many years have passed since then,
The flames of war have died,
The individual paths we chose
Are scattered far and wide,
But we remember proudly those
Whose lives to ours were tied,
When we were young.

Audrey Grealy

In 1943, future plans to bring the Pacific War to a successful conclusion rested on an Allied onslaught against the Japanese mainland by long-range bombers. Britain's contribution would be

'Life was a fleeting moment when / We lived from day to day / A morning dawned, the sun broke through / We savoured every ray . . .' (Author)

'Tiger Force'. However, following the dropping of atomic bombs on Hiroshima and Nagasaki by US B-29s on 3 and 6 August 1945, Japan surrendered on 15 August and there was no need to proceed with 'Tiger Force'.

'In war, resolution; in defeat, defiance; in victory, magnanimity and peace, goodwill.'
Winston Churchill.

Within three years, though, the western democracies faced a new threat both in Europe and the Far East.

PLAIN FARE

In 1948 the Soviet Eastern Bloc denied the Western Powers access to the divided city of Berlin, where the population of over 2.5 million was still recovering from the ravages of war and depended solely on surface transport links with the outside world. When the Soviets closed all the border routes on 25 June 1948, Berlin was totally isolated. Medicines and foodstuffs, materials and goods, could only be supplied to the beleaguered capital by air. On 26 June the Berlin Airlift began when the USAF started Operation Vittles with a flight from Frankfurt by a C-54 Skymaster. Two days later the RAF began Operation Plain fare. By 1949 crews from the air forces of Australia, New Zealand, and South Africa had joined the airlift, and by now an 8,000-ton day was not uncommon. The Russians finally caved in on 11 May 1949, and the Berlin blockade was over. In 10½ months, in 195,530 round trips, the Allies had delivered 1,583,686 short tons of supplies, as well as 160,000 tons of material to build or improve airfields.

'SABRE SONG'

Break Left,
Break Right,
Streamers on the wing.
Flick Roll,
Slow Roll,
We do anything.

'Berlin was totally isolated . . . and could only be supplied by . . . air' (using aircraft such as the Avro York seen here). (BAe)

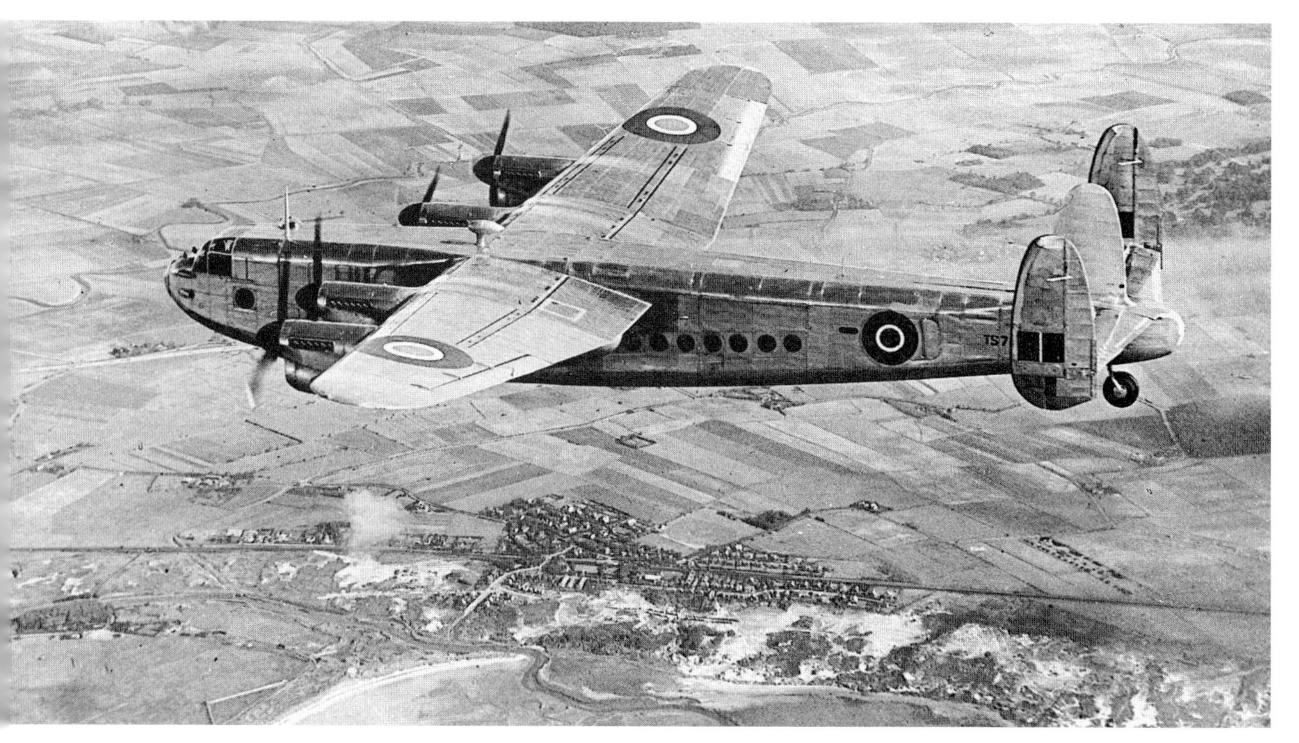

RAF Participation in the Korean War, 25 June 1950–1953

No 88 Squadron, Hong Kong
No 205 Squadron, Seletar
No 209 Squadron, Seletar
Far East Flying Boat Wing* (Short Sunderlands)

*Flew 1,100 in-theatre operations from Iwakuuni, Honshu, Japan, for the loss of one Sunderland, killing 14 crew and passengers.

- 21 RAF pilots operated with USAF fighter squadrons
- 29 RAF pilots served with No. 77 Squadron, RAAF
- 5 MiG-15s were confirmed shot down by five RAF pilots
- 10 RAF pilots were lost flying while attached to the USAF/RAAF
- 1 RAF pilot was lost flying with the FAA

"Another Reserve officer casually stepping out on to a wing that ain't there any more."

(Punch)

Sabre song. Some 430 US F-86 Sabres, licence-built in Canada and acquired under the Mutual Defence Aid Program, began equipping RAF Germany in 1953 and Fighter Command. By June 1956 all Sabres in Germany had been replaced by the Hawker Hunter. (Brian Pymm)

Far East Flying Boat Wing (Short Sunderlands), flew 1,100 in-theatre operations from Iwakuuni, Honshu, Japan, during the Korean War. (Author's collection)

BRUSHFIRE WARS

O ruler of the earth and sky, be with our airmen when they fly,
And keep them in thy loving care amid the perils of the air.
O let our cry come unto thee
For those who fly o'er land and sea.

In 1950 V-bombers were a distant dream and Avro Lincolns still bridged the bomber gap. The RAF's lean equipment programme was never better demonstrated than at the Farnborough Air Show in July of that year, when no fewer than 15 Lincoln squadrons performed a flypast. Although Washingtons (mothballed ex-USAF Boeing B-29s) would equip eight squadrons in Bomber Command from mid-1950 to September 1951, the English Electric Canberra, the RAF's first jet bomber, would not enter service until May 1951. Lincolns, and mostly piston-engined RAF fighters, fighter-bombers and transports, were used to help fight 'brushfire' wars in the Middle East, Africa, and the Far East. The Middle East saw the RAF involved in several actions during 1952–1967, and it was fully stretched to maintain peace and calm in the Trucial Oman, 1952–1959; Kuwait, 1961 (when Iraq laid claim to the oil-rich state); and Aden, where a fierce guerrilla war was fought in the Radfan area from December 1963 until 7 November 1967, after a rebellion had broken out. At the height of the troubles, in April-June 1964, Hunter FGA.9s alone flew 642 sorties.

In October 1952 Britain sent RAF squadrons and troop reinforcements from Egypt to Kenya to help put down insurrection by Mau-Mau terrorists. The Mau-Mau were a secret society among the Kikuyu tribe. In November 1953 part of 49 Squadron was detached to Shallufa, Egypt, and part to Eastleigh (Nairobi) Airport. Both

In 1950 V-bombers were a distant dream, and Lincolns still bridged the bomber gap. (BAe)

elements were reunited at Eastleigh late in November and remained there until January 1954, when a 100 Squadron detachment began three months of operations. On one occasion two of 49 Squadron's Lincolns, nine Harvards and police spotter aeroplanes took part in an operation in which almost ten tons of bombs were dropped on the terrorists. They were also strafed before British troops of the 39th Brigade began mopping up operations. In March 1954 a detachment of 61 Squadron Lincolns arrived at Eastleigh (Nairobi) and made repeated strikes on the Mau-Mau until June of that year. Another Lincoln squadron which saw action in Kenya in 1954 was 214, which had received Lincolns in 1950. By December 1956, when the emergency ended, the Mau-Mau had lost 10,572 killed and 2,633 captured.

OPERATION MUSKETEER

In 1956 426 RAF and French aircraft, based on Cyprus and Malta, were used to attack Egyptian airfields in Operation Musketeer from 31 October to 5 November, after President Nasser had nationalised the Suez Canal.

'In July 1956 Bomber Command was ill-prepared to undertake a Musketeer-type operation . . . The Command was geared to a "radar" war in Western Europe and was not constituted nor organized for major overseas operations . . . The majority of the Valiant force had neither Navigation Bombing Systems nor visual bomb-sights and were not cleared for HE stores . . . the Canberra aircraft forming the bulk of the force deployed are equipped only with *Gee-H* as a blind bombing device and it was not possible to position ground-based beacons to give coverage for this equipment over Egypt . . . It was considered that it would be prudent for the early attacks to be made at night and this necessitated a reversion to the marking technique successfully used in WWII . . .' *Musketeer Report.*

'The looks and expressions of surprise can only be imagined when, within two hours of landing at Luqa, all crews gathered in the Bomber Wing Operations briefing room for the first operational briefing and the curtains were drawn aside to reveal Egyptian airfields as the targets. Targets in Phase 1 were the Egyptian airfields operating Russian-built Il-28 bombers and MiG-15

Operation Musketeer, 31 October–5 November 1956

6, 8 & 249 Sqns	49 Venom FB.4	Akrotiri
13 Sqn	Canberra PR.7	Akrotiri
39 Sqn	Meteor NF.13	Akrotiri
208 Sqn	Meteor PR.10	Akrotiri
1 & 34 Sqns	24 Hunter F.5	Nicosia
10, 15, 18, 27, 44 & 61 Sqns	48 Canberra B.2	Nicosia
139 Sqn*	11 Canberra B.6	Nicosia
73 Sqn	Venom FB.1	Nicosia
70, 90 & 511 Sqns	16 Hastings	Tymbou
30, 84 & 114 Sqns	31 Valetta	Tymbou
9, 12 & 101 Sqns	22 Canberra B.6	Hal Far
109 Sqn	7 Canberra B.6	Luqa
138, 148, 207 & 214 Sqns	24 Valiant B.1	Luqa

Total sorties flown: 131 from Malta, 264 from Cyprus. 942 tons of bombs dropped. One aircraft (Venom) lost. (*target marking)

In Aden, a fierce guerrilla war was fought in the Radfan area from December 1963 to 7 November 1967. At the height of the troubles, during April-June 1964, Hunter FGA.9s alone flew 642 sorties. (Hawker Siddeley)

fighters, and at dusk on 30 October operations commenced. Airfields attacked by the squadron were Almaza near Cairo (on 30 October) and

Above left ***Kenya made world headlines in October 1952 when Britain sent RAF squadrons (a 204 Squadron Shackleton is shown at Eastleigh, Nairobi) and troop reinforcements from Egypt to put down insurrection by Mau-Mau terrorists. (Colin Walsh)***

Left ***During Operation Musketeer, Valiants and English Electric Canberra bombers (the latter being the first RAF jet bombers in history, May 1951) operated from Malta. (Vickers)***

Abu Sueir (on 31 October). The aiming points were the runway intersections, and crews were briefed to avoid the camp areas. Further instructions were given that bombs were not to be jettisoned 'live' in case Egyptian casualties were caused.'

Flying Officer R.A.C. Ellicott, 214 (Valiant) Squadron.

'No. 148 became the first V-force squadron to take part in operations by leading the attack against Almaza airfield by five Valiants and one from 214, four Canberras of 109 Squadron and three of No. 12 operating from Malta. The visual marking was done by Canberras [of 139 Squadron] operating from Cyprus. Canberras from Cyprus also carried out bombing on the same target. Little opposition was encountered. There was light flak around the target area, but it was sporadic and well below the attacking aircraft . . . Intelligence reports stated that there were ten Vampires, ten MiG-15s, ten Il-28s,

nine Meteors and 31 twin-engined transports on the airfield . . .'
148 Squadron diarist, 30 October 1956.

'On 31 October six crews were briefed to carry out an attack on Cairo West airfield. Led by the squadron commander, these crews took off during the afternoon, but after roughly one hour's flight all aircraft were recalled, as it was believed that American civilians were being evacuated by air from that airfield. After burning-off fuel, all aircraft brought back their bombs and landed safely. Later that evening two crews, captained by Sqn Ldrs Wilson and Collins, carried out an attack on the airfield at Abu Sueir. Both crews dropped proximity markers with 11,000 lb of bombs, and both marking and bombing were observed to be extremely accurate. No enemy opposition, either by fighters or anti-aircraft fire, was encountered by either crew.'
138 (Valiant) Squadron diarist, 31 October 1956.
(About 60 Egyptian aircraft were destroyed on the ground, but airfields were not put out of action, and all remaining aircraft fled south out of range and took no further part.)

FIREDOG

There was trouble in the Far East, too. In 1945 the defeat of Japan had created a power vacuum in the Far East which the emerging communist factions were quick to exploit. When the Malayan People's Anti-Japanese Army (MPAJA) was disbanded in December 1945 and told to hand in all weapons, some 4,000 refused and took their weapons into hiding. In June 1948 Emergency Regulations were passed and the guerrillas were forced into the jungle, where a vicious confrontation broke out between the British and Malayan Security Forces and Chin Peng's Malayan Races' Liberation Army (MRLA). Operation Firedog, the air war against the communist terrorists (CTs) during the Malayan Emergency, began in earnest in July 1948 with the formation of an RAF Task Force at Kuala Lumpur. Reinforcements, in the form of aircraft, notably Lincolns from the UK, were sent to Malaya and Singapore on detachment. The Malaya Emergency officially ended in August 1960. Firedog finished in October.

'The bomb aimer had the luxury of a seat in the new faceted nose compartment. The aircraft had white-painted upper surfaces to reflect heat when used in the tropics, and underneath was painted black. Ground crews regarded the Lincoln with

'The sincerest tribute I can pay the Lincoln after having such a tried and tested friend in the Lancaster is to quote the squadron motto:* Corpus Non Animum Muto *(change my body not my spirit).' (BAe via Harry Holmes)

affection from the point of view of servicing and maintenance because the aircraft was a definite improvement on the Lancaster; the motor cooling side sections were hinged to fall down and provide servicing platforms.

'I have one clear remembrance of a test flight on 9 September – the considerable flexing of the wings. I read later that under loading destruction tests the wing flexed 8 ft at the tips before failing – what a relief! From memory the stall was very gentle, with no tendency to drop a wing, and recovery was immediate. With flaps and undercarriage down, the stall occurred around 65 kt IAS. Crews had no difficulty in converting to the Lincoln because it was an aircraft that had no vices and was a pleasure to handle, but it was understandable that pilots who had extensive experience of the Lancaster were unwilling to concede it pride of place to any other aircraft, even a bigger and better Avro machine. The sincerest tribute I can pay the Lincoln, after having such a tried and tested friend in the Lancaster, is to quote the squadron motto: '*Corpus non Animum Muto*' (I Change My Body Not My Spirit).

Flight Lieutenant Frank Jones, commander of the Lincoln Flight at East Kirkby and Scampton, early September 1945 to January 1946.

All of the techniques employed in Malaya during 1948–60, and the lessons learned, were put to good use in 1963–66 in the confrontation with Indonesia after President Sukarno had opposed the creation of a Federation of Malaysia (which was proclaimed on 16 September 1963). Far East Air Force Hastings C.1/2/4 and Valetta C.1 transports did sterling work. Twin Pioneers, Beverleys and Hastings airlifted troops to Borneo. Beverleys played a key role in a 'hearts and minds' campaign in Brunei. Hunter FGA.9s of 20 Squadron, and later Javelin FAW.9s of 64 Squadron, were used to deter Indonesian violations over Malaysia. In May 1965 Hunters, directed by a forward air controller, routed the invaders with rocket fire.

One of the undoubted successes in the role of jungle transport and casevac was the Scottish Aviation Pioneer, which was used to supply the jungle forts in Borneo. Its excellent STOL characteristics were used to great effect by the pilots of 209 Squadron.

On 11 August 1966 the Confrontation (and the undeclared war in Borneo) ended. Britain and her Commonwealth allies had won a great victory in a terrorist war fought in the swamps of Malaya and jungles of Borneo against a strong, well equipped guerrilla force. Although as such the RAF did not mount a concentrated air war, the Confrontation certainly could not have been won without air transport. The victor was rewarded with a dismantling of its transport fleet during a gradual British withdrawal from the Far East in the late 1960s and early 1970s.

The withdrawal of British forces from the Far East involved a rapid reinforcement commitment, and, in the largest air-to-air refuelling operation then undertaken, in January

'The Twin Pioneer was "The Queen of the Skies".' (Flt Lt Rupert Aubrey-Cound)

Ten Lightnings were flown to Singapore and back, a distance of 18,500 miles, refuelled by Victor tankers. (Ministry of Defence)

1969 ten Lightnings were flown to Singapore and back, a distance of 18,500 miles, refuelled by Victor tankers. Departing from the UK in early January 1969, the Lightnings were each refuelled 13 times by the Victors, and 228 individual refuelling contacts were made, during which 166,000 Imp gallons of fuel were transferred.

'Some declared that flying the Pioneer was like "trying to fly an umbrella in a wind storm," but the Twin Pioneer was "the Queen of the Skies". Others claimed that, by comparison, the Beverley was like "flying a castle from the battlements". The Argosy, supposed to replace the Beverley, was nicknamed The "aluminium confidence trick," the rumour being that a smaller petrol bowser had to be designed to fit inside it. The Valetta, said to be difficult to land smoothly, was operated by 48 and 52 Squadrons, whom we called "The Ton-Up Boys".

'The Twin Pioneer was operated from improvised grass airstrips normally built across the bend of a river. This was usually the only piece of flat ground available in the rocky terrain. We operated on three-week detachments to Labuan and Kuching from Singapore, the single-engined Pioneers of "B" Flight flying into strips like Long Banga, Partick, and Liu Matu, which were too short for the Twin Pioneers of "A" Flight which I flew.'

Flight Lieutenant Rupert Aubrey-Cound, pilot, 209 Squadron, Borneo. (He also piloted Twin Pioneers in Oman during the Kuwait crisis.)

COLD WAR

Meanwhile, to counter the Soviet threat in Europe, on 4 April 1949 the North Atlantic Treaty Organisation (NATO) was formed, comprising Great Britain and the western democracies. That same year the Warsaw Pact came into being, embracing Russian and its satellites. The so-called 'Cold War', which was to last until the early 1990s, resulted in the RAF in the UK and Germany being placed on permanent alert in anticipation of Soviet and WarPac incursions.

THE GODS OF WAR

In 1946 the Air Ministry issued Specification B.35/46, calling for a strategic jet bomber capable of carrying a 10,000 lb bomb at 500 kts over a still-air range of 3,350 nm, with a ceiling of 50,000 ft over the target. A year later the British Government took the decision to produce nuclear weapons. At the time it was generally

Vulcans (a 617 Squadron Vulcan is shown) and Victors continued to provide a British nuclear deterrent until 1969, when the Polaris fleet became operational. (Hawker Siddeley)

accepted that high-speed, high-altitude jet bombers could outfly interceptors of the day. The Avro Type 698 (Vulcan) and the Handley Page HP.80 (Victor), the two winning designs, and the Vickers Type 660 (Valiant), adopted in 1948, were put into production, but protracted development meant that the Vickers design was the last of the three V-bombers to enter service. A total of 50 Victor B.1s were ordered for the RAF, and the type entered service with 232 Operational Conversion Unit (OCU) at Gaydon, Warwickshire, on 28 November 1957. In April 1958 10 Squadron at Cottesmore, Lincolnshire, became the first Victor operational unit.

By the start of the 1960s the appearance of the surface-to-air missile forced urgent modifications on the V-Bomber fleet, but by May 1965 the Valiant force was scrapped after metal fatigue problems were discovered. The Victor tanker replaced the Valiant tanker in the flight-refuelling and reconnaissance roles.

In September 1963 Victors of 139 Squadron became operational with the British air-launched Blue Steel stand-off thermonuclear missile. This weapon later equipped 100 Squadron Victors, as well as three squadrons of Vulcans. Plans to replace Blue Steel with the American Skybolt intermediate-range ballistic missile were thwarted when it was cancelled, but agreement was reached whereby the USA would supply Britain with Polaris missiles for launch from Royal Navy nuclear submarines. Vulcans and Victors continued to provide a British nuclear deterrent until 1969, when the Polaris fleet became operational.

AIR POWER

'NATO's deterrent posture cannot depend solely on deterrence provided by powerful conventional forces. Because those forces, good as they are, are numerically inferior to those of the Warsaw Pact in Europe, and because the deterrent effect of conventional forces has historically never been very certain, NATO continues to need nuclear forces as well. The awesome destructive power of nuclear weapons deters aggression in a way that tanks do not. Today the RAF's contribution in this area is sizeable – with British free-fall nuclear bombs for our Tornados and Buccaneers, together with the ability of our Nimrods to carry American depth bombs against submarines. The INF treaty – a triumph as it is for NATO's strategy of negotiating from strength – will, however, put an even greater premium on aircraft-delivered nuclear weapons, now that the INF surface-to-surface missiles on both sides are to be destroyed. Air power will therefore remain vital in this area.

'Because of NATO's continuing and essential reliance on nuclear weapons, their destruction – with conventional weapons alone – would be a vital Soviet war aim. As NATO's nuclear arsenal in Europe could be attacked essentially only from the air, air defence is crucial. In this area, surface-to-air missiles – useful as they are – are not enough, as the Falklands War showed. Indeed, given the continuing huge investment in the Soviet Air Forces, air defence fighter aircraft will remain an essential component of NATO's air defences for many years to come – hence our large investment in UK air defences today, and hence our plans to introduce the European Fighter Aircraft in the mid to late 1990s.

'But the denial of NATO's airspace to the Warsaw Pact cannot be achieved only by defensive aircraft and surface-to-air missiles.

'All our studies continue to show that the most effective way of achieving this aim is to disrupt enemy airfields – in much the same way as the Luftwaffe attacked RAF airfields in 1940 before it so foolishly switched its attacks to British cities. Our Tornado GR.1s with the JP233 weapon system are without doubt the best airfield attack combination in the world today. They provide SACEUR with an important part of NATO's offensive counter-air capability – another essential role for air power, if NATO is not to be defeated.

'The features that allow air power to carry out these vital tasks derive essentially from speed, reach, and the capability to deliver heavy fire power in a very short time. It is these qualities that allow air forces to carry out missions that could be accomplished efficiently in no other way. Air power remains an indispensable part of NATO's deterrent posture and the RAF makes an essential contribution to it.'

'Air Power – Our Raison d'être', Gp Capt Marten Van der Veen, 1989.

'The cancellation of the F-111K was met with some disappointment by the work up personnel. An informal cocktail party was organised by the Mess Committee, who started the proceedings by greeting the guests with the record "Hello Goodbye" by the Beatles.'

RAF Honington 1968. Work up personnel on the station had been told that the on-off purchase of 50 F-111K bombers, ordered to fill the gap left by the cancellation of the TSR-2, had suffered the same fate. Ironically, its replacement was the Blackburn Buccaneer, a fast and formidable low-level strike aircraft which had been offered to the RAF in the early 1960s, but had been rejected in favour of the TSR-2 and then the F-111. The "Brick", as it was universally known, was destined to play a major role.

THE 'BLACK LADY OF ESPIONAGE'

The Lockheed U-2 flew for the first time on 1 August 1955. It was to see service with both the USAF and RAF, and is best remembered for its clandestine spying missions over the Communist Bloc during the 1960s. Four RAF pilots, Michael Bradley, David Dowling, John McArthur, and Christopher Walker, were trained at Laughlin AFB, Texas, to fly the U-2. Then they returned to England, and to RAF Watton, Norfolk, to join the 'IO-IO' Squadron and take part in U-2 flights over the Soviet Union. RAF participation in Operation Overflight was a conditional part of the agreement which allowed American U-2 spyplanes to be based in East Anglia.

CORPORATE

'On 2 April 1982, shortly after the introduction of the Nimrod MR.2, Argentinian forces invaded the Falklands. The UN attempted to persuade the ruling junta in Buenos Aires that it should withdraw its forces of occupation from the islands, but to no avail. Even though the UK Government had already decided to mount an operation to retake the islands, Britain's action was legitimised by the UN following this failure to resolve the situation diplomatically.

'RAF maritime forces were soon despatched under Operation Corporate to Gibraltar and Ascension Island to provide surveillance along the naval force's route. Once established on Ascension, enhanced with an air-to-air refuelling (AAR) capability and fitted with the Harpoon (air-to-ground missile) and Sidewinder air-to-air missile, the Nimrods undertook daily security patrols around Ascension to guard against surprise attack from the sea. In addition to other

tasks, the Nimrods were also ordered to intercept, where possible, any vessels known to be attempting to defy the arms blockade imposed by the UN upon Argentina. Upon successful completion of Operation Corporate, the Nimrods were withdrawn.'
Squadron Leader Barry Wallace, 120 Squadron Nimrod captain.

The Falklands Conflict was the Harrier's conflict. The world's first operational fixed-wing vertical/short take-off or landing (V/STOL) aircraft began life as the Hawker P.1127, which hovered in the air for the first time on 21 October 1960, using a vectored-thrust turbofan. On 13 March 1961 the P.1127 Kestrel, as it was then called, made its first flight. The type was then evaluated by a British, American, and West German composite squadron at RAF West Raynham, Norfolk, from 15 October 1964 to 30 November 1965. These tripartite trials proved that the Kestrel was ideally suited to military purposes, and led ultimately to orders for 135 Harriers. The first of six development Harriers made its maiden flight on 31 August 1966, and the first true production aircraft flew on 28 December 1967.

In April-June 1982 Harrier GR Mk 3s of 1 Squadron played their part in the Falklands Conflict alongside Sea Harriers of the Royal Navy. While No. 1 Squadron's Harriers were fitted with AIM-9 Sidewinders (previously cleared for the AV-8As of the US Marine Corps and RN Sea Harriers only), on 2 May Flt Lt Paul Barton of 1 Squadron shot down the first Argentinian aircraft of the conflict, a Mirage III, flying a Sea Harrier of 800 Squadron. The first of 1 Squadron's GR.3s flew a non-stop 9 hr 15 min flight from St Mawgan, Cornwall, to Wideawake Airfield on Ascension Island on 3 May. This record distance of 4,600 miles beat the previous record held by a single-engine V/STOL aircraft of 3,500 miles from London to New York, set during the May 1969 *Daily Mail* Transatlantic Air Race. By 5 May a further seven Harriers had arrived on the island or were en route. Four Harriers were retained for the air defence of Ascension Island.

On 6 May six Harriers were embarked on deck of the SS *Atlantic Conveyor*, with eight Sea Harriers of 809 Squadron and four Chinook helicopters of 18 Squadron, and ferried to HMS *Hermes*, where, on 18 May, they made their first deck landings on an aircraft carrier at sea. On their first training sortie Wg Cdr Peter Squire, 1 Squadron's CO, and Flt Lt Jeff Glover intercepted an Argentine military Boeing 707 on a reconnaissance mission some 200 miles north-east of the Task Force. On the afternoon of 20 May Squire and his flight commanders, Sqn Ldr's Bob Iveson and Jeremy Pook, made an attack on a fuel dump at Fox Bay, on West Falkland, with cluster bombs. Next day Pook and Flt Lt Mark

The Falklands Conflict was the Harrier's conflict (seen here exercising in Norway). (BAe)

Hare attacked enemy helicopters in the Mount Kent area. They destroyed two Pumas and a Chinook on the ground before Hare was hit by several rounds and they returned to the carrier. Meanwhile, Squire and Glover provided close air support for the British amphibious assault in San Carlos water. Glover was hit by ground fire, ejected, and was taken prisoner. Later that day Flt Lt John Leeming and Lt Clive Morell RN destroyed two A-4 Skyhawks near Goose Green.

On 22 May 1 Squadron's Harriers attacked Goose Green airfield, and next day a Puma crashed as a result of their presence and two helicopters were destroyed. On 24 May four of 1 Squadron's Harriers, together with Sea Harriers, attacked Port Stanley airfield with 1,000 lb bombs. Next day they bombed the airfield again, and on 26 May the GR.3s made seven ground-support sorties in support of British troops advancing on Port Stanley. On 27 May, when the Harriers operated in support of the paratroops' advance on Goose Green, Bob Iveson was hit by 35 mm fire and downed. He ejected and evaded capture before returning to his unit. On 29 May Jeremy Pook also had to eject after being hit.

On 1 June Flt Lts Murdo Macleod and Mike Beech landed their replacement Harriers on *Hermes* after an 8½ hr flight directly from Ascension Island to the war zone, bringing 1 Squadron's strength up to five. On 5 June, when Harriers operated from near Port San Carlos, Jeff Glover was shot down and taken prisoner. Two more Harriers arrived, and on 9 June 1 Squadron flew four sorties against Argentinian gun emplacements on Sapper Hill and Mount Longdon. The next day ten ground attack sorties and a toss-bombing attack were made on Argentinian positions. On 12 June Murdo MacLeod was hit by small-arms fire but was able to land back on *Hermes.* On 13 June Harrier GR.3s made the first Paveway laser guided bomb attack in the Falklands with successful pinpoint attacks on two targets just before the Argentinian surrender was signed. The Harrier GR.3s had flown 126 operational sorties, including support missions for the landings at San Carlos, Darwin, and Goose Green.

On 30 April/1 May two Vulcans of 101 Squadron, one flown by Sqn Ldr Martin Withers and Flt Lt Dick Russell, co-pilot (pictured in 1994 climbing into one of the ten Victors that supported 'Black Buck', set out on a demanding 7,860-mile round trip to bomb Port Stanley in an incident filled, complex operation. (Author)

'The nearest operating base available to the RAF was Wideawake airfield on the British-owned island of Ascension, 3,375 miles from the Falklands, and Victors were needed to bridge the gap. The entire 55 Squadron effort was devoted to Operation Corporate. The first RAF bombing raid, codenamed "Black Buck", was made on 1 May 1982. The outbound leg involved a "snake climb" of 13 aircraft: 11 Victors and 2 Vulcans. The Vulcan successfully reached the Falklands and dropped 21 1,000 lb bombs on Port Stanley airfield. "Black Buck 1" had been the longest-range bombing mission in history.'

Wing Commander David Williams, 55 Squadron.

In April 1982 Victor K.2 tankers of 55 and 57 Squadrons at RAF Marham, Norfolk, were deployed to Wideawake on Ascension Island for operations in the South Atlantic. Beginning on 20 April, Victors, supported by five more operating in the air refuelling mode, flew three maritime reconnaissance operations, each of more than 14 hrs duration, to waters in the region of South Georgia.

On 30 April/1 May two Vulcans of 101 Squadron, supported by ten Victors, set out on a demanding 7,860-mile round trip to bomb Port Stanley in an incident-filled, complex operation. At the start the primary Vulcan aborted after its cabin could not be pressurized, and the Victor fleet, too, suffered malfunctions. Victors refuelled Victors until only two remained with the reserve Vulcan, flown by Flt Lt Martin Withers and Flt Lt Dick Russell. Just before the fifth refuelling of the Vulcan one of the Victors attempted to refuel the other Victor in very turbulent conditions, and the receiving Victor had its probe broken during the transfer. The two Victors exchanged roles and the provider took back the fuel. Although dangerously low on fuel itself, the Victor then transferred enough fuel to the Vulcan to allow it to make its attack, before heading back towards Ascension and calling for another tanker to meet it. Withers' successfully dropped his 21 1,000 lb bombs on Port Stanley airfield. The Vulcan was then refuelled a sixth time, and returned to Ascension after being airborne for 15 hrs 45 min. At the time it was the longest-range operational bombing operation ever flown.

Between 30 April and 12 June five "Black Buck" raids by Vulcan bombers, using up to 16 tankers, were made on Argentinian targets. The final bombing operation involved a Vulcan and 14 Victor tankers which carried out 18 refuelling sorties.

'The whole of the back of the aircraft was literally boiling oil fire into the sky . . . I told the chaps it'd be a good idea to stop as quickly as possible and then we'd all get out and run as fast as you can.' (Author)

'It was not long after the Falklands Conflict when I was scrambled early one morning with my Victor crew to go and support a fighter. We'd just started the take-off roll when there was a muffled thump. There were no adverse engine indications. My captain said: "Did anybody hear that?", as if to say: "Has your knap bag fallen over?" I said I'd heard it. He then made the best decision of his life. He decided to abort. He closed down the throttles and started slowing down. I got out the periscope to view the undersurface of the aircraft and discovered that we were massively on fire because one of the engines had blown up and the debris had cut through the fuel tanks. The whole of the back of the aircraft was literally boiling oil fire into the sky . . . I told the chaps it'd be a good idea to stop as quickly as possible and then we'd all get out and run as fast as we can. Have you ever tried running in an immersion suit, flying boots, and a Mae West over a thick grass airfield? It's jolly difficult.'

Wing Commander Al Beedie, OC 55 Squadron, RAF Marham.

SKY GUARDIANS

O Trinity of love and grace, true guide of all who fly through space,
In peace and war, midst friend or foe, be with them whereso'er they go.

'During peacetime, unannounced and uninvited Soviet reconnaissance aircraft frequently pass through and operate with the UK ADR (Air Defence Region). Aircraft of No. 11 Group are charged with identifying and shadowing these aircraft within the UK ADR. Within minutes of an "intruder" being detected, one or two fighters can be winging their way towards the visitor –

not in anger, but as what may be termed the 11 Group UK ADR "escort service". The fighters will observe behaviour of the Russians, collect visual and photographic reminders of their visit – for later evaluation by intelligence officers – and ensure that they eventually leave the ADR. This role is carried out unceasingly by the QRA, or Quick Reaction Alert Force, 24 hours a day, 365 days a year, in fair weather or foul for an average of some 200 incursions annually.

'The first indication of approaching trade is often provided by Norwegian radar stations, as the Russians head west from their home bases . . . In our scenario, it appears that one or possibly two Soviet aircraft are heading south-west towards the UK ADR . . . The aircraft have been positively identified as Bear Delta maritime reconnaissance aircraft, door numbers 19 and 35 – old friends and regular customers of the escort service. The Russian pilots are no fools, though, and on this occasion, understandably wishing to go about their business in private, they descend to low level and fade from the radar screens. Faeroes Radar, which was tracking them, now loses contact but assumes they are still heading our way.

'In our scenario, it appears that one or possibly two Soviet aircraft are heading south-west towards the UK ADR . . . The aircraft have been positively identified as Bear Delta Maritime reconnaissance aircraft . . .' (USAF)

'Meanwhile, a Phantom from Leuchars is en route to its Combat Air Patrol position under control of Buchan and then Saxa Vord. A VC 10 tanker has been diverted from an exercise in the North Sea, and is heading north with two Tornado F.3 interceptors from the Southern QRA at Coningsby. US forces in Iceland have now reacted following discussion with Buchan and have launched AWACS [airborne warning and control system] and two F-15s. A Shackleton is heading for his barrier patrol to attempt to locate the low-flying Bears, and control of these assets is passed to Saxa Vord.

'An hour later the Shackleton has picked up an intermittent contact, and directs the fighter towards it. Soon, the contact appears on Saxa Radar and a controlled intercept can begin. Until the fighter can pick up the target on its own radar, it devolves upon the skill of the intercept controller at Saxa Vord to move the fighter towards its moving target in a huge game of three-dimensional chess. Rather hoping that their flying petrol station is in the vicinity, the Phantom crew are soon able to pick up the Bear on radar and move in to escort. The Bear pair split at low level and now that this has been confirmed, the Tornados, having already tanked, are moved further west towards the Americans who have now got a contact from the AWACS. Saxa Vord passes them on to the control of the US forces and a mixed group of F-15s and Tornados is directed to the second intercept of the day.

'Although the Bears have been picked up again they are not unduly worried, as it gives them an opportunity to study Western fighters at close range and for them also to take snaps for the boys back home. But now they've got better things to do and they continue towards the edge of the ADR. The Americans pull off at this stage and control of the Tornados and F.4 passes to Benbecula. The VC 10 is still there and the combine moves slowly south-west. The Phantom departs for home as the Russians leave the ADR still heading south-west, but the two Tornados and the tanker remain with them until certain that the Bears are not about to reverse course.

'The Bears are now moving into the UK Southern Sector of operations controlled at Neatishead in Norfolk and its associated Radar stations. Bishops Court in Ulster and then

'. . . heading north . . . Tornado F.3 interceptors from the Southern QRA at Coningsby. US forces in Iceland have now reacted following discussion with Buchan, and have launched AWACS . . .' (Sergeant Rick Brewell)

Portreath in Cornwall monitor the Bears' progress as they head towards the Bay of Biscay to go about their legitimate business.

'The operation you have just read about is typical of a day in the life of No. 11 Group AD units. It does get very much busier at times, but the system copes. The changes which are now under way will revolutionize our AD capability. Air Defence is definitely here to stay.'
Group Captain Nick Buckley, 1989.

GRAND GESTURE

You who fly Tornados from off Maharraq's Sands
We think you are doing marvellous, we think you're doing grand;
You made us proud we're British, sons of that old breed,
Which down the years of history, provides when comes the need,
Gallant men, our nation's pride, to face our every foe;
To stand like at Kohima or like you, go in low.
Yes, very low, and mighty! and take the lethal hail!
To leave the target useless as it passes 'neath your tail
It's doubtful if you notice on your nocturnal ride
That there's a vast armada, flying at your side
If you could use your inner sight, I'm sure you'd see
Planes that flew before you, E'en in the RFC
The Blenheims and the Battles which also faced the muck
Who, all too often, saw their mates running out of luck,
Lumbering old Whitleys, Wimpys, Halleys, Lancs
Assorted wood Mosquitos, some with long-range tanks
And see those queer old biplanes that know that land so well?
They flew it all between the wars, and have their tales to tell

'. . . all the Victor assets . . . were needed to deliver the Tornado GR.1s, F.3s, and Jaguars from Europe to Tabuk, Dharan, and Muharraq.' (Mike Rondot)

So you in those Tornados? You'll never fly alone!
Old bomber boys are with you! They make a mighty drone
And as you fly your sorties, let no one dare deride
All, who are proud they're British in thought are by your side
PS: Forget old chairborne generals, in dotage, slightly barmy.
They probably attained that rank whilst in the Sally Army.

Jasper Miles

'In August 1990, 55 Squadron was supporting RAF Jaguars at the Reconnaissance Air Meet in Texas when the recall of all Victors to the UK was ordered. Within 24 hours the tankers were back at Marham, and within 48 hours they were operating over France and Sicily to deploy fast jets to the Gulf. Numerous tasks, involving all the Victor assets, were needed to deliver the Tornado GR.1s, F.3s and Jaguars from Europe to Tabuk, Dharan, and Muharraq. On completion there was a formal handover from 101 Squadron VC 10 detachment, which had been operating at the base for three months. The next day three more Victors arrived, shortly followed by two more crews. The initial requirement from Strike Command was that the Victor detachment should support the Tornado F.3 and the Jaguar missions only, and the VC 10 detachment would support all Tornado GR.1 sorties . . .'

Wing Commander David Williams, CO, 55 Squadron during Operation Grandby.

'Few would dispute that the Gulf Conflict, Britain's biggest military campaign since World War II, was highly successful . . . within 48 hours of the British political announcement, there was a squadron of Tornado F.3 air defence fighters at Dhahran in eastern Saudi Arabia; two hours later they were flying combat air patrols protecting the integrity of Saudi airspace; 48 hours later, a squadron of Jaguar fighter-bombers arrived at Thumrait in Oman with their vital VC 10 tanker support. Shortly thereafter, Nimrod maritime patrol aircraft started their essential sea surveillance operations in support of the UN embargo in supplies reaching Iraq; and these activities were backed up by a huge logistical operation which got under way with incredible speed and efficiency. Within four days the RAF had deployed a powerful force of offensive, defensive and combat-support capabilities over 3,000 miles and brought it rapidly up to full operational status . . .'

Group Captain N.R. Irving AFC RAF, writing in 1992 about Operation Granby, the RAF operation in the Gulf War, 1991.

'I knew that in the early stages of any conflict at least, the Tornados would primarily be used for night operations. This meant that most of my 24 crews were allocated to the night shift. As the UN deadline loomed I thought that it was

RAF Dispositions in the Gulf War, 1990–91

SAUDI ARABIA		
Tabuk	19 Tornado GR.1	
Riyadh (King Khalid Int)	7 Hercules	1 TriStar KC.1
	9 VC 10 K.2/3	1 HS.125
Field Locations	17 Chinook HC.1	12 RN Sea King
	19 Puma HC.1	
Dhahran	18 Tornado F.3	6 Tornado GR.1A
	13 Tornado GR.1	
BAHRAIN		
Muharraq, Bahrain	13 Tornado GR.1	12 Buccaneer S.2
	12 Jaguar GR.1	7 Victor K.2
OMAN		
Seed	2 Nimrod MR.2	

ELEMENTS OF THE FOLLOWING SQUADRONS DEPLOYED:	
Tornado F.3: 5, 11, 23, 25, 29 & 43 Sqns	(18 aircraft)
Tornado GR.1: 9, 14, XV, 16, 17, 20, 27, 31 & 617 Sqns	(45 aircraft)
Tornado GR.1a: 2 & 13 Sqns	(6 aircraft)
Buccaneer S.2: 12 & 208 Sqns	(12 aircraft)
Nimrod MR.2: 42, 120, 201 & 206 Sqns	(4 aircraft)
Chinook & Puma: 7, 18, 33 & 230 Sqns	(17/19 aircraft)
C-130 Hercules: 24, 30, 47 & 70 Sqns	(7 aircraft)
Jaguar GR.1: 6, 41 & 54 Sqns	(12 aircraft)
VC 10 K.2/3: 101 Sqn	
TriStar K.1: 216 Sqn	(17 aircraft)
Victor K.2: 55 Sqn	
HS.125	(1 aircraft)

'There was a squadron of Tornado F.3 ADF at Dhahran . . . two hours later they were flying CAP protecting the integrity of Saudi airspace . . .' (Ministry of Defence)

prudent to initiate the night shift system so that our body clocks would adjust to the conditions. Then, in the afternoon of 16 January, whilst the night shift was sleeping, I got *the* 'phone call

from Gp Capt Dave Henderson, the Bahrain Detachment Commander. He said: "Bring everybody in".

"In?" I queried.

"Yes". "Brief everybody."

'In the first week of January I had lost the 12 Tornado crews who had the most theatre experience when 14 Squadron went home. (Air crews changed over every six weeks in the Gulf). In their place I got six from 27 Squadron and six from 617 Squadron. 14 Squadron's ground crews, who were on a three-month detachment,

'A squadron of Jaguar fighter-bombers arrived at Thumrait in Oman with their vital VC 10 tanker' (seen here with Jaguars and Tornado GR.1 ZD791/BG, shot down on 17 January 1991 when it was being flown by Flt Lts Peters and Nicholl). (Ministry of Defence)

In the first months of the operation the tankers were VC 10s of 101 Squadron. These aircraft have the capability to allow two Tornados to tank at the same time – one from each of the VC 10's wing-mounted hoses. (Flt Lt Chris Carder)

Tornado GR.1A ZA372/E* Sally T*, based at Dhahran, Saudi Arabia. (RAF Marham PR)

stayed! XV Squadron, my squadron, plus attrition reserves from 9 Squadron, made up the rest of the Muharraq detachment. I was in charge of elements of five squadrons from three air bases in two commands; RAF-G and Strike!

'I told the crews to rest, but how could they now! There were some really serious looking young men around. We had all hoped it wouldn't come to this, that instead Saddam would come to his senses and back down. This was not our back yard. We weren't fighting for the White Cliffs of Dover. It's very difficult in the cold light of day in a distant land to fight for a principle. None of our people would choose to die in the desert, given the option, but their attitude was: "Let's get it done, as professionally

'Nearing the target, Nigel [Sqn Ldr Risdale] said: There's a bloody great punch-up going on . . ." Wing Commander John Broadbent. (Mike Rondot)

as we can, and let's get home as soon as possible". Ours was arguably the most dangerous mission going in the Gulf. Nevertheless, while there was a military reason for what they were being asked to do, my people were determined to execute it to the best of their ability. Equally, I was determined too, and knowing what the overall picture was, I knew I had a good chance of arranging it.

'I put XV Squadron crews into the first two waves, 27 and 617 Squadrons would cover the following night. Squadron Leader Nigel Risdale and I would lead the first eight Tornados. Our target was [Al] Tallil airfield in southwest Iraq, a huge airfield twice the size of Heathrow. We were to cut the two 11,000 ft parallel runways, and the access taxiways leading from the HAS's [hardened aircraft shelters] with our JP233 anti-airfield weapons. [Each contains 30 runway cratering submunitions and 215 area-denial submunitions. Wing Commander Jerry Witts of 31 Squadron led four Tornados from Dhahran to the same target].

'It was a funny feeling, emptying the contents of your pockets into little plastic bags to be collected by the pretty Squadron Intelligence Officer. It was an odd moment when she gave us

'We were to cut the two 11,000 ft parallel runways, and taxiways leading from the HASs, with our JP233 anti-airfield weapons.' (Ministry of Defence)

our "Goolie Chits", evasion maps, and gold sovereigns before we trooped off. We gave ourselves plenty of time to prepare. Very few people in the detachment knew we were going to war at this stage. Not even the ground crews knew. We'd carried JP233 on night training missions before, so this appeared no different. We tried to make it look like a normal training mission. It worked.

'It was very lonely taxying out with radio and lights out. Group Captain David Henderson and his OC Ops, Ray Horwood, gave us a brief wave. Our all-up weight of 30 tonnes with drop tanks and two JP233s (alone weighing 10,000 lb) meant we could not taxi to the into-wind runway. The net result was a very short taxi followed by a tailwind take-off.

'All eight of us got airborne, checked in with AWACs, and met up with our VC 10 tankers. After tanking at around 15,000 ft we dropped down to 500 ft crossing the Iraqi border. Then we went down to 200 ft as we approached the target area. On the ingress frequency we heard the Weasels and F-15s as they checked in. I could not of course see them in the darkness, but it was reassuring to hear them. The American package of about 30 aircraft were going to the same target and would bomb from medium altitude (around 20,000 ft). We were to bash through one minute after them, at 04:08 local, so that our JP233 minefield would be undisturbed.

'About 70 miles from the target Nigel said: "There's a bloody great punch-up going on," or words to that effect. He had his NVGs [night-vision goggles] on, whereas I didn't. I looked up and just saw darkness. Glancing at my watch, it was 11 o'clock – 4 o'clock local, the time the first American aircraft was due over Tallil. It was obvious the firefight was coming from Tallil and that we'd have to fly through it. I asked: "Just how fast can this Tornado go?" Nigel hit the burners, then took the burners out before we came over the visual horizon of the target. The firefight was right on our nose. Triple-A fire was hosing the sky, an occasional HARM [high-speed anti-radiation missile] was going down, and SAMs [surface-to-air missiles] were going up. It was all just incredible.

'My wave of four were converging on the airfield on a heading of 343°; the rear four on a heading of 078°. There would be one Tornado over target every 20 sec, so that after 2 min 20 sec all eight would have delivered their JP233s at low level and cleared the target at 550 kts. We rushed over the target wings level for 5–6 sec to get the JP233s away. There was a bloody great wall of flak. I motored my seat down so it could not distract me. In peacetime our cockpit seats would have been high. It was like a pretty firework display. Balls of tracer and cannon fire were coming towards us. It was like trying to run through a shower without getting wet. All underneath the aircraft started to light up. The empty containers slung under the fuselage automatically jettisoned. Then, we were out. "Holy **** – I've survived," I thought. Exhilaration! I felt ten foot tall – if not taller. Our attack was perfect.

'I waited a couple of minutes before checking the rest of the formation in. All responded until I got to No. 5 – nothing. "5 Check". Still nothing. "We've lost 'Gordo' (Gordon Buckley), leader of the back four, I thought. "We've lost him. I'm not surprised. Bloody good mate. Paddy Teakle too. Two really good hands. Never see them again." My exhilaration was gone . . . In our

'We rushed over the target, wings level, for 5–6 sec to get the JP233s away . . .' (Hunting Engineering)

Tornado clangers were going off WAH, WAH, WAH, red lights were flashing, and "Kojaks" (warning sirens) were blaring away. We were not going as fast as we should. We suddenly realized we had lost an engine. In the heat of battle we had not realized it. We did not jettison our tanks, and in any case the aircraft was much lighter without the JP233s. We retraced our steps across the desert and turned back to the tanker. We were pretty quiet because we thought we had lost "Buckers". Then, climbing up to the tanker, we heard him! He had heard us calling him but he had a weak transmitter and we could not hear him!

'We landed back at Muharraq at six in the dawn sky, some 4hr and 5 min after take-off. The world's press was waiting. We were first back. All eight Tornados were there. Rupert Clark, my own No. 4, told the press – "It went on rails!"

'The guys couldn't have done better. I couldn't have asked them to do more. (Post-strike photos showed a very successful outcome and no more than a 20 ft error). There was lots

'"We've been hit! We've been hit!", John Nichol [left], my navigator, shouted from the back seat . . . "Prepare to eject!" [Flt Lt Peters, right] yelled . . .' (Author and RAF Marham PR)

of "Biggles banter" with the troops. Squadron Leader Pablo Mason and the guys waiting to go out in the next wave [to Ar Rumaylah airfield] looked grey. We said the usual, that it was "a piece of piss," "nothing to worry about", etc! I rushed into squadron HQ still wearing my G-bags. Throwing open Dave Henderson's door I shouted: "We're all back!"

'In the first three days I lost three crews. It turned from being a jolly jape to a lot more serious. But nobody "wibbled". They just got on with it. If we'd had losses like that with the WARPAC, we'd have been delighted.'

Wing Commander John Broadbent, OC Tornado Detachment, Muharraq, who led a record 21 war missions from Bahrain, three of them JP233 sorties. After air superiority had been gained after just four days, the Tornados changed to medium-level bombing using dumb bombs and then, from 2 February, laser-guided bombs. He was awarded the DSO.

'The Tornado was doing 540 kt 50 feet above the desert when the missile hit. A hand-held SAM-16, its infra-red warhead locked inexorably on to the furnace heat of the aircraft's engines. Some lone Iraqi's lucky day. Travelling at twice the speed of sound, the SAM streaked into the bomber's tailpipe, piercing the heart of its right turbine. Five kilograms of titanium-laced high-explosive vaporised on impact, smashing the thirty-ton aircraft sideways. It shuddered, a bright flame spurting from its skin; fifteen-million-pounds-worth of high technology crippled in a moment by the modern equivalent of the catapult. The computerised fly-by-wire system went down, transforming the aircraft instantly from thoroughbred to carthorse.

'We had just completed our attack run on the huge Ar Rumaylah airfield complex in southwestern Iraq; I was pulling the Tornado through a hard 4g turn, with 60° of bank, to get on to the escape heading. The aircraft was standing on one wing, at the limits of controllable flight. The fly-by-wire loss sent it tumbling, the stick falling dead in my hands – a terrifying feeling for a pilot. I was pushing the controls frantically, the Tornado falling out of the sky, the ground ballooning up sickeningly in my windshield. The huge juddering force of the blast had knocked the wind out of me. Gasping, hanging off the seat straps, I yelled: "What the hell was that?"

"We've been hit! We've been hit!" John Nichol, my navigator, shouted from the back seat. Urgently, he transmitted to the formation leader: "We've been hit! We're on fire! Stand by."

"Prepare to eject, prepare to eject!" I yelled. "I can't hold it . . ."'

Flight Lieutenant John Peters, XV Squadron Tornado GR.1 pilot, shot down on 17 January 1991. From Tornado Down *(Michael Joseph, 1992). Peters and Nichol ejected, and were captured and later tortured before being freed by Iraq.*

RAF Losses, Operation Granby, 1990–1991

1990

18 Oct – Tornado GR.1 ZA466/FH. Sqn Ldrs Ivor Walker & Bobby Anderson, 16 Sqn. Crew ejected following a take-off collision at Tabuk with wrongly raised arrester barrier.

13 Nov – Jaguar GR.1A XX754. Flt Lt Keith Collister+, 54 Sqn.

1991 (all Tornado GR.1s)

13 Jan – ZD718/BH. Flt Lts Kieran Duffy/ Norman Dent++, 14 Sqdn.

17 Jan – ZD791/BG. Flt Lt John Peters*/Flt Lt Adrian 'John' Nicholl*, XV Sqn. Hit by AAA en route to Shaibah/Basrah.

17/18 Jan M – ZA392/EK. Wg Cdr Nigel Elsdon°/Flt Lt Max Collier°, 27 Sqn. Hit by SAM after attack on Ar Rumaylah air base.

18/19 Jan – ZA396. Flt Lts Dave Waddington*/Robbie Stewart*, 27 Sqn. Downed by a Euromissile Roland SAM.

20 Jan – ZD893/AG. Sqn Ldr Peter Battson/ Wg Cdr Mike Heath, 20 Sqn. Control malfunction on take-off. Crew ejected.

22 Jan – ZA467/FF. Sqn Ldrs Garry Lennox°/ Paul Weeks°, 16 Sqn. KIA during attack on Ar Rutbah radar station.

23 Jan – ZA403/CO. Pt Off Simon Burgess*/ Sqn Ldr Bob Ankerson*, 17 Sqn. Premature explosion of one of their own bombs.

14 Feb – ZD717. Flt Lts Rupert Clark*/ Stephen Hicks°, XV Sqn. Hit by two SA-2 SAMs on LGB attack over Al Taqaddum air base.

+Killed in training, Oman/++ Saudi Arabia. *PoW. °KIA.

Total Sorties flown by RAF: 4,000 combat/ 2,500 support. Bombs dropped: over 3,000 tonnes. 262 aircraft committed (162 at any one time). Personnel committed (all British services) 45,000. Total cost to GB (all forces) £2.4 billion.

'We were only flying over Iraq for about 30 minutes until we hit the target. Absolutely pitch

'My exact words were: "*! Missile!" I broke left and shouted: "Chaff!" at Robbie.' Robbie Stewart, centre, arm raised, and Flt Lt Dave Waddington, left, on their return to RAF Marham. (RAF Marham PR)***

dark, nothing around us – just like doing a training sortie. There were eight aircraft in the formation. We were carrying 1,000 lb bombs, trying to toss them into the airfield to suppress the triple-A so that the last four aircraft, who were carrying JP233s, would have an easier time.

'We were just coming up to the pull-up point where we would release the bombs, three-and-a-half miles from the target. I remember the missiles being fired at me, seeing it at 12 o'clock,

Tornado GR.1 ZA466/FH, from which Sqn Ldrs Ivor Walker and Bobby Anderson of 16 Squadron ejected following a collision with a wrongly raised arrester barrier on take-off from Tabuk. (RAF Marham PR)

'The aircraft never let us down, and the now-famous female "art forms" that adorned them are testament to the affection that they inspired.' (Author)

Tornado ZA392/EK, which was hit by a SAM after the attack on Ar Rumaylah air base on 17/18 January. Wing Commander Nigel Elsdon, OC 27 Squadron, and Flt Lt Max Collier were killed. (RAF Marham PR)

which is the worst position. My exact words were: "****! Missile!" I broke left and shouted: "Chaff!" at Robbie. All I could see was a flame like a very large firework coming towards me, but once I banked the aircraft we lost sight of it. Then there was an enormous wind – I think I was unconscious very quickly. My last thoughts were that I was going to die.'

Flight Lieutenant Dave Waddington, 27 Squadron Tornado GR.1 pilot. He and his navigator, Robbie Stewart, were shot down by a Euromissile Roland SAM on 18/19 January 1991, during the mission to Al-Tallil air base near Basra, captured, and made PoW.

'As we progressed through the various phases of the war, from night low-level JP233 attacks to medium-level operations and finally to laser-guided bomb deliveries in concert with the worthy Buccaneers, we came to realize what a very good aircraft the Tornado is. Certainly, the aircraft never let us down, and the now famous female "art forms" that adorned them are testament to the affection that they inspired, and we did ask an awful lot of them. The aircrew averaged 20 sorties or so apiece during the six-week campaign. The aircraft averaged about twice that number and, no matter what anyone tells you, machines do get tired after a while.'
Wing Commander Jerry J. Witts DSO, MBIM, 31 Squadron, OC Tornados, Dhahran.

VICTORS OF THE OLIVE TRAIL

'January 1991, additional Victor aircraft plus two of the initial crews were positioned at Muharraq and, by 16 January, 55 Squadron had in theatre a total of six aircraft, eight crews, and 99 groundcrew. Further training sorties were flown until 16 January when, at 22:50 hrs, two Victors led the first Muharraq Tornado GR.1 bombing mission into Iraq. The sortie was flown along the Olive Low Trail, which was a track running generally south of the Iraq border but concluding with a short northerly leg, which cast off the receivers into the heart of enemy territory. Olive Trail then became the bread-and-butter route for the rest of the war. To meet all the contingencies and to ensure that fuel was available for the Tornados on their return from the mission, all Victor aircraft were refuelled to the maximum 123,000 lb for take-off. The early sorties were affected by the very poor weather along the refuelling tracks and, consequently, the aircraft encountered three times the normal usage of fatigue. As experience grew, the take-off fuel was adjusted and the fatigue penalty was reduced.

'On 19 January an additional Victor was deployed to the Gulf, as up to a maximum of 14 sorties were being flown per day. Many missions were flown over the Persian Gulf in support of attack aircraft and air defence patrols and,

'Two Victors led the first Muharraq Tornado GR.1 bombing mission into Iraq . . .' (Sergeant Rick Brewell)

'The Victor detachment achieved every objective' (Mike Rondot)

together with 138 Olive Trails and numerous other combat air patrols, 299 sorties were flown over the 42-day war, an average of 33 missions per crew. The Victor detachment achieved every objective and did not fail to complete a single task or sortie. It was tight at times, but flexibility, excellent engineering support and good airmanship saved the day and produced 100 per cent success rate.'

Wing Commander David Williams, CO 55 Squadron during Operation Grandby.

'The Gulf was the Victor's final curtain call . . .' (Author)

The Olive Low Trail was used throughout the 42-day war, in which the Victors not only refuelled aircraft from Britain, but also from Canada and France, and those from the USA that were probe-and-drogue compatible.

'I thought that with the current situation out in the Gulf it would be nice to paint my own nose art on the Victors. I'd got the idea from the Second World War American nose art on Fortresses like *Memphis Belle*. I approached the boss and he said, "Go for it". Each aircraft was assigned a crew chief in the Gulf and each Victor was named after the crew chief's wife or girlfriend. So they're lucky guys in a way if they look like that!"
Corporal Andy Price, 55 Squadron.

The Gulf was the Victor's final curtain call, although there were many who believed that, with continued care and maintenance, the Victor could have been kept in service well into 1994 and possibly beyond.

'. . . The sheer scale of the airlift operation increased the cargo throughput at RAF Brize Norton tenfold. This enormous task exceeded even the Berlin Airlift. During the whole of Operation Granby, 45,759 personnel and 52,661 tons of freight were airlifted to the Gulf. From November 1991 additional civil charter was used to assist in clearing the backlog at the height of the airlift. This involved 38 Belfast and 28 USAF C-5 Galaxy charters, which accounted for 69 per cent of the total airlift capacity. Considerable recourse was made to sealift, and the bulk of the RAF's weapon stocks were transported in this way.'
Wing Commander Dil Williams, OC Engineering and Supply Wing, RAF Marham.

'The Operation Granby land offensive lasted only 100 hours, but its success and the massive effort required to prepare, train, and support a force of such a size over such huge distances is a spectacular story. From its arrival in theatre in January 1991 to the achievement of its final objective astride the Basrah Road on 28 February, 1 (UK) Armoured Division travelled the equivalent of the distance from the Normandy Beaches to Berlin. For a fighting force of 25,000 men, 523 armoured vehicles, 84 artillery pieces, and over 1,000 wheeled vehicles,

'. . . final objective astride the Basrah Road.' Fleeing Iraqis caught on the 'Road to Hell' from Kuwait City to Basrah, 28 February. (Ministry of Defence)

'18 crews and 12 "desert pink" Buccaneers deployed on 27 January 1991.' (Mike Rondot)

not including four-tonners and Land Rovers, it was clearly a major undertaking. The Support Helicopter Force played its part in this build-up and it too was faced with logistic and resupply problems well in excess of anything ever planned or achieved before.

'Over 5,000 hr were flown by the three types [Puma, Chinook, and Sea King], during which 2,750,000 lb of freight were carried, 17,500 troops lifted, and over 300 casualties evacuated for emergency treatment. This was achieved largely from desert sites – 31 in all, not including Signals detachments – and during a deployment across the desert of 1,000 miles over difficult and mainly hostile terrain . . .'
Wing Commander R.E. Best AFC RAF.

BANANA JET

Despite a long operational life of some 30 years, the Blackburn Buccaneer, prior to the Gulf War, had never seen 'active service'. However, continual updating of the aircraft's avionic and weapons capabilities ensured that when the RAF required laser designation support for its Tornado aircraft during Operation Granby, the

Maritime Buccaneer Wing of RAF Lossiemouth was suitably equipped and prepared for combat. Eighteen crews and twelve 'desert pink' Buccaneers deployed on 27 January 1991. Operational flying began on 2 February, when the first of many targets in Iraq was attacked using Paveway LGBs, delivered by Tornado and guided by the Pavespike laser designation system of the Buccaneer.

'On all missions two Buccaneers would accompany four Tornados to their allocated target. After long transits to Iraq, tanking en route, the package would split into two Tornados and one Buccaneer. The Tornados would approach the target and release their 1,000 lb Paveway guided bombs. The Buccaneer crew would then be responsible for guiding these bombs to the target.

'The necessity of the Buccaneer pilot visually acquiring each target from high level meant that meticulous study and planning for every target took many hours. Tornado crews, as the leaders of each package, took the brunt of most of the planning, with the Buccaneer crews concentrating on the actual attack details.

'Planning began the day prior to a mission, and crews usually reported for briefing six hours prior to take-off. Co-ordination with AWACS, tanker support, ECM [electronic counter-measures] support aircraft and fighter aircraft all had to be arranged before sortie briefing, with lots of time left afterwards for intelligence and combat survival updates. Most operational missions involved long transits before entering enemy airspace; this time was gratefully used to ensure that all aircraft systems were synchronised with the accompanying Tornados that they were all working.

'Once into enemy territory everything focused on completing the task. All of one's senses were heightened, but as the target approached one seemed to get tunnel vision and nothing mattered more than finding the aiming point and marking it until the bombs impacted. The Pavespike pod encases a television camera which gives the Buccaneer navigator a small picture on a screen in the rear cockpit. Once his pilot has pointed the camera at the target the navigator can watch the aiming point and fire the laser beam at it. Seconds after "bomb gone", although it often seemed like an eternity, the Buccaneer crew were able to evaluate the success of their attack and have the results recorded on a video tape.

'Now the pilot and navigator could revert to being a crew again, and all eyes were out of the cockpit, looking for enemy surface-to-air missiles and anti-aircraft-artillery (AAA), as all haste was made towards friendly territory.'

Squadron Leader W.N. Brown DFC, 'A' Flight Commander, 12 Squadron.

'. . . the RAF's capability for the airborne designation of LGBs . . . provided by the Pavespike pod fitted to the Buccaneer . . . was a 20-year-old design and was fitted with only a TV camera. Hence it could provide only a daylight capability. The necessity for night operations during Operation Granby meant that TIALD (thermal imaging airborne laser designator) would be rushed into service as soon as possible. The only TIALD pods available at the time were two flight demonstrator pods with TI only, which had been flying on a Buccaneer testbed aircraft. These had never been designed for carriage on the Tornado; indeed, there were no Tornado aircraft capable of carrying the pods either. However, a rapid development programme, TAP (Tornado advanced programme) was undertaken at the Aeroplane & Armament Experimental Establishment (A&AEE), Boscombe Down, and the two pods, affectionately called Sandra and Tracy, entered operational service on 10 February 1991, flying from Tabuk in Saudi Arabia and destroying hardened aircraft shelters at the H3 south-west airfield complex in north-west Iraq . . . As testament to their outstanding success, the two pods flew 91 missions in 18 days, scoring 229 direct hits. Overall their success rate was bettered only by the F-117A Stealth Fighter.'

Flight Lieutenant A.S. Frost BSc, navigator, 617 Squadron, who carried out the first operational TIALD sortie during Operation Granby.

DESERT CATS

'We are on 30-minute alert for CAS [Close Air Support]. We do not expect to be launched. Then from our tasking authority in Riyadh a flash message. They have a mission for us to Iraq, time over target (TOT) two hours. Mayhem. Maps are found, thoughts are gathered, then the automatic routine takes over. We are ready to go, final brief from the GLO (Ground Liaison Officer) –

Squadron Leader Dave Bagshaw AFC, at 54 the oldest pilot in the Gulf War, flew 23 missions, including five bombing operations. His Jaguar, XX733, was complete with pink Spitfire caricature. (Author and Steve Jefferson)

people wish us luck, we in turn try and smile, not really showing how we feel. The engineers have prepared the aircraft. More good wishes. A final walk around the jet, all the pins are out. A final nervous chat with the groundcrew helping me to strap in – then a smile and he is gone and I am on my own. I go through the mission in my mind, reminding myself of the lessons the pilots learnt yesterday and telling myself that I won't make the same mistakes. Almost before I am ready it is time to start engines – no problems here, as usual – the Jaguar is fully serviceable.

'Mary Rose' by Chris Frome on Jaguar XZ356 flown by Wg Cdr Bill Pixton, CO 41 Squadron, RAF Coltishall. (Mike Rondot)

Check-in time arrived, everyone is on frequency, serviceable and ready to go.

'Once airborne I carry out operational checks to make sure all the aircraft systems are working. The ECM pod self-tests, lighting up the radar warning receiver (RWR), chaff is fired from the

'We used inflight refuelling on about half the missions, always with the Victors. They were brilliant, they were always there, they never failed to turn up, they were always in the right place, at the right time, with the right fuel.' (Mike Rondot)

Phimat chaff dispenser and the flares come out from the ALE40 flare dispenser. I check-in the formation with the AWACs; lots of friendly voices with the best information from our controller being "Picture clear" – no Iraqi aircraft airborne. We continue northwards, climbing to high level. As we approach the border we are handed over to our assigned ground control agency. He tells us the weather is poor in the target area, but is able to give us an alternative target in Kuwait where the weather is suitable for our mission. The weather improves as we cross the border. ECM pod is in automatic mode, weapon switches are made live and the RWR starts to light up, indicating all manner of systems, some friendly, some not. I take in some of this information, then the target area approaches – weapon aiming selected, find the target and into the dive. I seem to be in the dive for an eternity, in reality only seconds – and then the weapons are released and I start to climb away. No time for proper battle damage assessment (BDA), but my bombs appear to have exploded in the area of the target. Back over the Gulf I have time to think clearly. I check-in the rest of the formation – they are all there, weapons delivered. In no time at all we are taxying in at Bahrain. Smiles, handshakes, and congratulations all round, nobody is thinking that we may have to do this again tomorrow. Back to the squadron operations room. Debrief the mission with the GLO, EWO, and IO [Ground Liaison Officer, Electronic Warfare Officer and Intelligence Officer] and then back to the hotel for the squadron nightly debrief.

'So ended my first day of operational flying.'
Squadron Leader C.M. Allam, 41 Squadron, Thursday 17 January 1991.

'Virtually every mission we flew had dedicated SAM suppression, and you felt like you had minders all the way, with air defence stuff roaming about, clearing the skies. We used in-flight refuelling on about half the missions, always with the Victors. They were brilliant, they were always there, they never failed to turn up, they were always in the right place, at the right time, with the right fuel. We just used to pitch up, get in behind them, take our fuel and go away, not a word being said on the radio unless they were in thick cloud. Even then we could get very close using air-to-air TACAN [tactical air navigation], but if you couldn't find them, you had to talk.

'We had all sorts of targets, mainly in Kuwait and southern Iraq. Barracks, storage depots, artillery, SAM sites, coastal defence missile sites, troop emplacements, installations on airfields, things like that. If you looked at the intelligence map of where we were operating, we were attacking targets in the area with the heaviest concentration of known Iraqi defences, so while we felt sorry for the guys flying deep-penetration missions into Iraq, or for the A-10s down at 8,000 ft over the open desert, looking for tanks, they had a lot of respect for what we were doing.'
Squadron Leader Mike Rondot, Flight Commander 6 Squadron, seconded to 41(F) Squadron: 29 Jaguar missions in the Gulf War.

JAGUAR PILOT'S NOTES

Step 1: Approach the aircraft in a steely-eyed, devil-may-care manner. This creates a favourable impression on spectators and groundcrew. Do not trip over any cables, etc., as this does not create a favourable impression.

Step 2: While conducting pre-flight checks, walk around the aircraft knowingly, peering intently for several seconds at any complicated arrangements (e.g. flaps, undercarriage, etc). This will fool your groundcrew into thinking you know what you are doing.

Step 3: Borrow a chinagraph pencil from your marshaller and jot down the date (your groundcrew should know). Ask the time and jot that down also. Check the tail code of your aircraft and . . .

Step 4: Proceed rapidly to your assigned aircraft and repeat steps 1–3, not forgetting to inform Wing Ops to delay your clearance by ten minutes. Ignore raucous laughs.

Step 5: Climb quickly into the cockpit and sit down. If you have the control column in your left hand and the throttle in your right, you are SITTING THE WRONG WAY ROUND. Rotate your body through 180° (a half turn). You should now be able to see your marshaller in front of you.

Step 6: Arrange all switches, levers, etc, in a

pleasing and eye-catching manner; at this point your groundcrew may start waving at you. If this occurs, rearrange all switches etc, until correct combination is obtained, whereupon they will stop waving.

Step 7: Ignite engines and, advancing all throttles, roll smoothly over the chocks, skilfully avoiding the saucer of milk in front of the nosewheel.

Step 8: Air Traffic Control will now pass instructions to prevent you getting lost on the taxiway. Upon reaching the end of the taxiway, apply brakes and run the engines up to 80 per cent power. This should provide skid marks approximately 50 yards long. NOTE: The thrust from the engines is a natural result of advancing the throttles, and is not to be described as 'engine surge' as is frequently reported.

Step 9: When lined up on the runway, close eyes, push throttles fully forward, and count up to ten before pulling the stick back. If contact with the ground has not been re-established after five seconds, open your eyes and continue with the mission.

Step 10: Call Wing Ops and ask them what your mission was.

'A final walk around the jet, all the pins are out. A final nervous chat with the ground crew helping me to strap in . . .' (Ministry of Defence)

'Just off target there was a lot of flak. It's the first time I have seen tracer coming up at me. It was the longest minute of my life, I can tell you.'
Jaguar pilot's first mission over Kuwait.

'The Iraqi chemical warfare threat was assessed as being based largely on well-known agents – mustard and nerve, and on conventional delivery systems: amongst others, artillery, rockets, bombs, and ballistic missiles. This latter capability was exemplified by the Al Hussein missile, an Iraqi derivative of the Scud. It had a range of up to 600 km and had the potential to deliver chemical agents against most Allied facilities in the theatre of operations. Another threat was from biological weapons (BW). Though the exact status of the programme could not be discovered, the threat remained. BW agents, with very high toxicity relative to weight and a large hazard area, had potential for strategic disruption. To counter these physical and psychological factors, the RAF mounted a comprehensive response. One

'The Gulf War . . . achievements are seemingly endless and were brought about by total commitment, great endeavour, and no small amount of personal courage . . .' Buster Gonad and Flt Lt Steve Thomas return to RAF Coltishall. (Steve Jefferson)

Jaguar XZ119* Katrina Jane *returned to Coltishall from the Gulf. (Author)

measure of its and similar Allied countermeasures' success was that the Iraqis did not use CB weapons . . . the RAF, like the other Services, had excellent individual equipments, most notably the Mark 14 NBC Suit, the S10 respirator with its built-in drinking device, and the Chemical Agent Monitor (CAM). The last advantage was concentration of executive responsibility for

low-level implementation of CB defence and training on the RAF Regiment.'
Squadron Leader A. Brown, 'NBC Defence'.

'The Gulf War could not have been won without air power. Within the grand coalition, the RAF's contribution was substantial. During the campaign the RAF deployed 158 aircraft and 7,000 personnel to the region. A total of 6,500 sorties was flown during the six-week war, of which 2,000 were offensive missions into Iraq and Kuwait by Tornado GR.1s and Jaguars. Our reconnaissance Tornados and support helicopters also played a vital role, whilst our air transport fleet flew 13 million miles carrying 25,000 personnel and 31,000 tonnes of freight. The achievements are seemingly endless, and were brought about by total commitment, great endeavour, and no small amount of personal courage. The end result was a highly effective air campaign which paved the way for a swift and decisive land offensive, proving conclusively the immense reach, flexibility and speed of reaction associated with air power.'
Air Marshal Sir Andrew Wilson KCB, AFC, RAF, C-in-C RAF Germany and formerly British Forces Commander Middle East.

PER ARDUA AD ASTRA?

*'Per Ardua ad Asbestos' (F*** you, I'm fireproof)*

'In the past three years, since the break-up of the Eastern Bloc, defence cuts have cost the RAF five strike attack squadrons, five squadrons of Nimrod maritime patrol aircraft, and three search-and-rescue squadrons. Eleven airfields have closed: two more will go in 1996.

'In the next four years the RAF's numbers will plummet from 93,000 to 52,000 . . . One senior officer told me: "The MoD has promised to protect the front line but it is already badly stretched. We have detachments in Italy, Saudi Arabia, Turkey, and the Falklands. There is too much to do and too few people . . ."'
John Gibb, Sunday Express Classic *magazine, 12 February 1995.*

'Whilst the RAF of 1995 may be vastly different from the RAF of 1945 in most respects, today's RAF still maintains the same fighting spirit and devotion to duty that characterized the Service's contribution to the Second World War and which helped to bring about final victory. Air power proved to be the dominant feature of that war, and in those six years air power grew from adolescence to maturity. The RAF of today has a proud tradition to maintain, a tradition which was forged in the skies over Germany and Burma, and which is continued in the 1990s in the skies over Iraq and Bosnia.'
Chris Hobson, Senior Librarian, RAF Staff College, Bracknell.

OPERATION WARDEN

'Operation Warden is the UK contribution to the US-led Operation Provide Comfort relief operation [for the UN "Safe Haven" for the Kurds of Northern Iraq, established in July 1991 following their failure to overthrow Saddam Hussein]. Nos 1(F), 3(F), and IV(AC) Squadrons are responsible for providing the UK Offensive Support Contribution, primarily tactical air reconnaissance, within the Area of Responsibility (AOR).

'No. 1(F) Squadron deploys to Incirlik, Turkey, on rotation with the other two Harrier GR.7 squadrons . . . Once deployed, the more senior pilots lead the first few sorties, but from then on the lead is rotated between all the pilots. Flying over Iraq, with the huge array of hostile SAM systems and AAA deployed on the ground, certainly concentrates the mind and improves your lookout. Even with the Harrier's excellent integral electronic warfare suite, the ZEUS system, all pilots are constantly looking out for that telltale plume of smoke from a SAM missile launch. An early "spot" could make the difference between flying home and walking home!

'As was shown by an incident that occurred on 24 November 1993, the Kurds are obviously very aware of our presence in the AOR and of its protective benefits. A Harrier pilot had to eject over the Kurdish area of Northern Iraq after his aircraft suffered an engine failure. The people of the village of Bishiel witnessed this and sent out

'No. 1(F) Squadron deploys to Incirlik, Turkey, on rotation with the other two Harrier GR.7 squadrons . . .' (BAe)

men to help the pilot. The Muktar of the village sent a handwritten letter in barely understandable English to the coalition forces. In thanks for the help given by the village people, they received a bountiful supply of livestock and equipment. The village must now be one of the most prosperous in the area.'

Flight Lieutenant C.R. Soffe BSc, 1(F) Squadron Harrier GR.7 pilot.

OPERATION JURAL

'Operation Jural began in August 1992. Air assets from France, the United States, and the

'Operation Jural began in August 1992. Air assets . . . in the form of Tornado GR.1s were sent to enfore a UN no-fly zone in southern Iraq . . .' (Ministry of Defence)

UK, in the form of Tornado GR.1s, were sent to enforce a UN no-fly zone in southern Iraq, south of the 32nd Parallel. Throughout the operation the RAF has had a dual role. Equipped with the Ferranti thermal imaging airborne laser designator pod, the Tornado force has been tasked with carrying out reconnaissance of Saddam Hussein's operations against the Marsh Arabs south of the 32nd Parallel, and with providing a capability to deliver LGBs on to point targets. This latter capability was ably demonstrated during the Gulf War and during the raids on Iraqi air defences in early 1993.

'The RAF contingent based in the Arabian peninsula must stand ready to repeat these operations at short notice should the situation require it. On almost every day since the beginning of the operation, RAF aircraft have been operating inside Iraq, south of the 32nd Parallel, gathering information on the deployment of Saddam Hussein's forces.

'The squadron was fortunate in having a number of aircrew with considerable combat experience in operation Desert Storm. Even so, flying deep into Iraq with known SAM systems deployed is not to be taken lightly. Enemy radars and triple-A systems are keen to play cat-and-mouse with coalition aircraft – sometimes to their cost, as the launching of a USAF anti-radar missile in support of one of our missions demonstrated.'

Flight Lieutenant D.J. Knowles BA, 9 Squadron, RAF Bruggen.

MARTOCK, VIGOUR AND CHESHIRE

'Where [there is] famine . . . or when warring factions use food – or lack of it – as a weapon, as in the case of Bosnia . . . inevitably the UN High Commission for Refugees (UNHCR) attempt to provide the innocent bystanders with the basic essentials for survival. Lacking any resources, they quite naturally ask individual nations to contribute aid. In the case of the UK that request

'During Operation Vigour, in three months two Hercules and four crews of 38 Group delivered some 3,500 tons of supplies to all areas of Somalia . . .' Hercules of 38 Group at Moi Airport in March 1993. (Author)

comes to the Foreign & Commonwealth Office, who, if they agree, will inevitably "contract" the MOD to do the task, who in turn order Strike Command, who pass it to 38 Group, which is where we come in.

'Some of the activities and people involved in 38 Group's response to humanitarian tasks [have included] three recent but very different operations; Operation Martock, the evacuation of British Nationals from Luanda in November 1992; Operation Vigour, the UK contribution to the US Operation Provide Relief, involving the delivery of aid to Somalia from December 1992 to March 1993; and Operation Cheshire, the ongoing provision of humanitarian aid to the people of Sarajevo. During Operation Vigour, in three months two Hercules and four crews of 38 Group delivered some 3,500 tons of supplies to all areas of Somalia, flying just short of 1,000 hrs in the process . . . Finally, I turn to Operation Cheshire, which must be the riskiest operation undertaken by any aircraft in the RAF at the moment. The UN asked nations to provide aircraft to deliver aid to Sarajevo. The UK Government responded with an offer of one Hercules aircraft, and it has been flying into Sarajevo three times a day since July 1992. Regrettably, our aircraft are regularly tracked by radar-layed AAA and occasionally pick up transmissions from potentially hostile systems. Clearly they are at their most vulnerable during approach and departure at Sarajevo; therefore the ground situation is continuously monitored.

'As at the end of October 1993 the RAF had delivered some 12,500 tons of aid to Sarajevo in some 880 visits, and flown close to 2,000 hrs in the process. This represents some 18 per cent of all aid delivered by air – not bad when we represent only 12 per cent of the aircraft dedicated to the airlift . . . We have been fortunate so far, in that there has been little damage to our aircraft – only two bullet holes. Others have been less so; the Italians lost a Fiat G.222 aircraft to missile fire in 1993, and it was only through the quick reaction of the crew that the Germans did not lose a C.160 in February 1994. The hazards are evident. Our hopes are that we shall continue to get the balance right and not exceed it. We get it wrong at our peril.'

Group Captain D.K.L. McDonnell OBE, RAF, Head of Air Transport and Air-to-Air Refuelling Branch, HQ No. 38 Group, High Wycombe.

In August 1992 a multi-national air operation known as 'Provide Relief' was launched to airlift supplies to feeding centres and clinics for Somali refugees. By 25 February 1993 some 28,050.86 tonnes of food had been delivered by the USAF, German Air Force, and Royal Air Force based at Moi International Airport, Mombasa, in 1,924 sorties to Somalia and 508 within Kenya.

OPERATION DENY FLIGHT

Operation Deny Flight, a large force drawn together by the UN and NATO to support peacekeeping operations in Bosnia-Herzegovina, began on 7 April 1993 with the deployment of RAF Jaguars, Tornado F.3s, E-3D AEW aircraft, TriStar tankers, and Nimrods to forward bases in Italy.

'The mission was slated for the small hours. After a week and a half of waiting, this was to be my first. The brief was a model of economy, the delivery of a monotone. It was not a pep talk. It was the full stop at the end of a professional's sentence, expressionless and matter-of-fact.

'I dressed in slow time, reluctantly collected a pistol and two clips of ammunition, and stood out on the porch, watching the moon rise above a thin veil of cloud. Over Bosnia, only scattered rafts of stratus interrupted the view. Glancing down at the lights in towns and villages, it seemed as quiet as any country marooned in the very depths of night. I reached for my night-vision goggles. NVGs are little image-intensifying binos that fit to the front of a modified helmet on a lever arm so that they can be swung down over the eyes. Whilst light to the hand, NVGs are still heavy to the head and, like most backseaters, I prefer to use them as a captain's spyglass. I switched them on and peered down at the quiet scene 20,000 ft below. Through the pebble-dashed liquid green light of the NVGs I looked down at the bright flare of street lamps, and for the first time saw the still brighter flares of fires.

'We shifted our combat air patrol to the south as the second pair split to the tanker, perched over the Adriatic. Overhead Sarajevo we witnessed a war in progress, coloured pea green; fires, explosions, tracer fire.

'For a moment I started, thinking I saw an

aircraft flying low and fast across the battle lines, but when it bloomed into a sudden flash of light I realized it was a shell I'd been watching. In my guidebook it says: "Sarajevo is a deeply attractive city where the region's three historical

Operation Deny Flight began on 7 April 1993 with the deployment of RAF Jaguars, Tornado F.3s, E-3D AEW aircraft, TriStar tankers and Nimrods to forward bases in Italy. (Flight Lieutenant Chris Carder)

'Sarajevo is a deeply attractive city where the region's three historical ingredients – nationalism and Turkish and Austrian occupations – underscore a buoyant individualism.' (Author)

ingredients – nationalism and Turkish and Austrian occupations – underscore a buoyant individualism". It's the target for any trip here. A buoyant individualism that made this last remark the quote of the decade.

'To the north, storm clouds were gathering, and our view of the battle below grew dim before their riding and colossal bulk. These were the real gods of war, the genuine article, wearing white, flickering with fire and light, they sang out their rage across a land consumed by rage. We skirted their edges, slipped through the gaps; we were inconsequential by their side. Storms like this one would still be sweeping the skies of Bosnia long after the last shells had been expended, the last bullets fired, the last homes torched, the last villages surrounded and mortared and cleansed, the last escaping victims turned to refugees. Long after Bosnia had teetered to the very edge of abandon and tumbled headlong into the abyss, such a storm as this would still be customary across the night skies, a summer storm brewed from the midday heat.

'The radar homing warning receiver lit up and my head was filled with the trilling alarm of a triple-A battery target tracking. The target was tracking us. I stared mesmerized at the screen, feeling nothing. Fumbling in the dark, I reached for the chaff button and pressed it. The green run light flashed four times. Still the alarm sounded inside my head.

"Are you going to call that?", asked my pilot impatiently. I pressed the foot switch.

"Three One Bravo, targeted, one two seven." The alarm died. The screen cleared. I flew my final mission six weeks later. The trip spanned a sparkling, clear-skied midday. After 70 hrs of live armed inactivity, we were finally vectored into low level to search for helicopters hidden near Tuzla, and almost missed them. Almost. I looked down the left side of the jet and saw a pair of "Hips" [Mil Mi-8s] nestling in a sand quarry, dressed with insulting prettiness in coats of blue-striped white paint.

'I do not know if finding those two helicopters served a purpose. I do not know if it helped a country stumble along the dark path to peace. I only know that I would like to believe it.'

Flight Lieutenant A.G. Tait BSc, AMI MechE, XI Squadron F.3 Tornado navigator.

'Although we had known for weeks that the squadron was likely to deploy to Italy to assist in the enforcement of a no-fly zone (NFZ) over Bosnia, political imperatives saw us delayed and delayed. In the end we had just 48 hrs in which to move from RAF Leeming and commence operations out of the Italian Air Base at Gioia Del Colle in southern Italy. We flew the Tornados out early one April morning, refuelling from VC 10 tankers over the French Alps. Whilst excited at what might be in prospect, none of us had much idea what the sorties over Bosnia would bring. Certainly none of us thought that our first missions in Operation Deny Flight would be flown in total darkness!

'For the first two weeks of operations XI(F) Squadron had drawn the short straw and was confined to the "death watch", as we called it. This required us to get airborne shortly after 01:00 in the morning for a mission that would normally last four to five hours. The missions were necessarily long and rather tense in the first few days of operations. But after a week or so they became relatively short affairs. Once airborne, two of the fighters would head straight for the operating area while the other two would join up with a tanker over the Adriatic. In the first months of the operation our tankers were VC 10s of 101 Squadron. These aircraft have the capability to allow two Tornados to tank at the same time – one from each of the VC 10's wing-mounted hoses. After two months the VC 10s were replaced by the much larger TriStar, which can refuel only one aircraft at a time, but does have a lot more fuel to give away.

'Although the night missions provided very little in the way of "trade", they had other things to keep us occupied. There would be the UN flights dropping relief supplies, and there was all the activity on the ground. During daylight the ground fighting was difficult to observe from our fighters, sitting at 20,000 ft or so. But at night we could use the night-vision goggles. Consequently, the fighting always seemed to be far more intense at night, probably because that was the only time we could really see it. One night we sat high over Sarajevo and watched three or four separate battles rage. Detached from the horror of it all, we felt we were watching a firework display. It was only later when we saw the news reports that the true level of carnage and suffering was realized.

'Eventually the squadron was tasked with some daylight missions. With these came the "trade". After many hours of patrolling over Bosnia we started to find some helicopters, mainly Russian-built Mi-8 "Hips", which were operating in defiance of the NFZ. We would then shadow the helicopters until they either landed or left the NFZ. If they did neither, then the fighters would work their way through the Rules of Engagement, trying to contact the helicopters on the international distress radio frequency. The helicopters were warned that they were in violation of a UN resolution that they should leave the NFZ, land, or risk being attacked by the fighters. In all cases during XI(F) Squadron's time supporting Operation Deny Flight, all helicopters intercepted by us either left the NFZ or landed.

'Had we been tasked to engage one of the aircraft we intercepted, the Tornado F.3 is armed with the Sky Flash radar-guided, medium-range air-to-air missile, the Sidewinder heat-seeking missile, and a 27mm cannon. These weapons can be employed either visually, through the pilot's head-up display, or non-visually, using the aircraft's radar. Prior to deploying to Italy the squadron had undergone a phase of training against RAF helicopters to validate tactics and to ensure all crews were acquainted with the

'Because of our NVG capability, the Tornado F.3s tended to be used more by night than by day, and so were soon back to the "death watch". The only advantage this shift had was that we often saw the dawn break over Bosnia.' (Ministry of Defence)

'The Squadron's third deployment took place on 12 October 1994, at a time when tension was running high in Bosnia . . .' A 6 Squadron Jaguar from RAF Coltishall flies over Bosnia. (RAF Coltishall)

problems of finding and then attacking a low, slow helicopter.

'Because of our NVG capability, the Tornado F.3s tended to be used more by night than by day, and so were soon back to the "death watch". The only advantage this shift had was that we often saw the dawn break over Bosnia. Having watched through the goggles the mayhem of a country tearing itself apart, it seemed unreal that everything could suddenly look so peaceful and picturesque. The wispy dawn mist that formed in the valleys, shrouding the villages that had been burning all night, gave the land a ghostly appearance against the brightening eastern sky which silhouetted distant thunderclouds and mountains. Turning south for our transit back to Gioia Del Colle, we would start to relax and try to stretch aching limbs in the cramped cockpit. As we crossed the Adriatic at 30,000 ft the sun would be just clearing the eastern horizon. This was just the first sunrise we would see, for our descent into Gioia would mean the sun disappeared below the horizon again, only to rise once more as we were taxying back to dispersal. By now, though, many of us were too tired to care about the glories of an Italian dawn. We just wanted to get to bed.'

Wing Commander J.A. Cliffe, 11 Squadron, RAF Leeming.

'The squadron's third deployment took place on 12 October 1994, at a time when tension was running high in Bosnia. Initially operations continued at a normal pace with a mixture of reconnaissance and CAS sorties. However, at the beginning of November the world's attention fell on the Bihac Pocket in the north-west of the country. The use of missiles in the surface-to-surface role against a UN protected area caused particular concern. But it was the attack on 19 November against the town itself by two Orao aircraft from Udbina airfield in the RSK that finally provoked a NATO air strike. On 21 November a force of over fifty aircraft was involved in a mission to strike the runways and taxiways on the airfield. Included in this package were four Jaguars, two dropping 1,000 lb bombs and two carrying out post-strike reconnaissance . . .'

Wing Commander Tim J. Kerss MBE, BSc, OC 54(F) Squadron.

'Funny thing, it was meant to be my day off. I was going to play golf (we did six days on, two days off). I got a telephone call at about 08:30, telling me to come in. At briefing we were told that in the afternoon we would be raiding Udbina and I was to lead the Jaguar post-attack recce pair. Basically, I was to get post-strike BDA (Bomb Damage Assessment) photos using the LOROP (long-range oblique pod) at a height above 15,000 ft, 4–5 miles off the target.

'We tanked from a TriStar on the way in. Tanker after tanker was stacked up to refuel the American F-16s and F-18s and French Mirages.

It was a gorgeous day. You could see for miles. AWACS cleared the package to push through, then we got specific clearance to go in. An F-18, relaying the clearance from AWACs to us, said: "You are cleared to press. 'Manpads' (man-portable shoulder launched missiles) and triple-A are in the area." If flak was heavy we could always get the pictures later, but there was only light triple-A. The raid went in and palls of smoke were rising from the airfield, which was at the base of a mountain. The wingy asked: "What speed are you going through?"

I said: "As fast as I can get."

'I couldn't look out that much. I had to maintain level height to get good pictures and was using the HUD [head-up display]. My back-up was there just to watch my six. That was it. I got the photos and we all headed back to Giola Del Colle. We didn't need the tanker on the way home.'

Flight Lieutenant Chris Carder, 54 Squadron Jaguar recce pilot.

'We tanked from a TriStar on the way in . . .' (Sergeant Rick Brewell)

'On 21 November a force of over 50 aircraft was involved in a mission to strike the runways and taxiways on the airfield. Included in this package were four Jaguars, two dropping with 1,000 lb bombs . . .' (via Flt Lt Chris Carder)

'. . . and two carrying out post-strike reconnaissance.' Five of the Jaguar crews who took part. (via Flt Lt Chris Carder)

AIR MOBILITY

'The main lift for the British Army's 24 Airmobile Brigade is provided by three RAF squadrons equipped with the 51 ft-long Boeing Vertol HC.1/2 Chinook and Aérospatiale Puma HC.1 helicopters. At RAF Odiham No. 7 Squadron has 18 Chinooks (and two Gazelle HT.3s), and 33 Squadron has 12 Puma HC.1 helicopters. At Laarbruch, Germany, 18 Squadron has 5 Puma HC.1, 6 Chinook HC.1 and one Gazelle HT.3. First delivered in November 1980, the Chinook is the largest helicopter in the RAF inventory. It can carry 28,000 lb of freight and operate at an all-up weight of 50,000 lb. Some 17 RAF Chinooks and 19 Pumas operated during the Gulf War.

'During Exercise Gryphon's Eye in May 1995, Pumas and Chinooks shuttled back and forth from STANTA [Stanford Training Area] to RAF Woodbridge, Suffolk, where HQ were quartered. One might think the RAF personnel would be enjoying all the luxury a permanent station affords, but not a bit of it. Everyone, including air crews, were wearing full NBC (nuclear, biological, and chemical) kit for anything up to 12 hrs at a stretch, and the headquarters staff were operating out of tents inside a hangar which was once home to American A-10 Thunderbolts. In the command tents white strip lights and bulbs painted blue and red, illuminated the map tables and sections marked LOGS, PLANS, CURRENT OPS, and TASKERS. The HQ was protected by 150 personnel of the RAF Regiment who were taking the defence of the station very seriously indeed. Puma crews were practising the whole gamut of NBC drill, carrying their PV (portable ventilators) to and from their helicopters like city gents with attaché cases, and going through the very uncomfortable ritual of dressing and undressing in the decontamination tent for rest periods in full NBC conditions.

'Once airborne in the Puma, again it was back to whizzing and weaving along the contours of the Suffolk countryside, punctuated only occasionally by the need to vault over the trailing power lines and pylons of the national grid. In the battle area again it meant darting behind the tree lines and skirting the hedgerows like a fighter pilot trying to evade a dogged yet imaginary pursuer.

'Out in the field at STANTA, Chinook and Puma crews were not under NBC conditions but under canvas nonetheless. Air and ground crews were putting in a customary 14-hour day in the Ulu (countryside). One of the most exacting elements of the whole exercise is for the Puma and Chinook pilots to fly at night in the battle area. For night read day. It is all made possible by the use of NVGs, which cost £16,000 a copy. A modern war is fought 24 hours a day, so a force that cannot operate during the hours of darkness is non-effective. NVGs have been used successfully in helicopters for many years now, and their use extended to day-only aircraft such as the Jaguar and Harrier. NVGs are fitted with light-intensifying tubes which amplify electromagnetic energy in the visible and near-infra-red part of the spectrum. The goggles, which actually look more like binoculars, have a field of view of 40° and are mounted in the pilots' and loadmasters' helmets. Despite a rather cumbersome appearance, they are not heavy and are quite comfortable to wear. Improved "Generation 3" tubes enable pilots to fly in much lower ambient light levels where no moonlight is required at all. Certain modifications to the aircraft lighting are required before NVG flying is possible. Normal cockpit lighting, being rich in red and near-infra-red wavelength light, is too bright for the NVGs. Therefore, to enable the pilot to see the outside scene through the NVGs unobscured by reflections, and view the cockpit instruments around the NVGs at normal light levels, cockpit lighting with blue-green filters is used.

'Throughout Gryphon's Eye, Chinooks and

'The main lift for the British Army's 24 Airmobile Brigade is provided by three RAF squadrons equipped with the 51 ft-long Boeing Vertol HC.1/2 Chinook.' (Author)

'Throughout Gryphon's Eye, Chinooks and Pumas use the West Tofts Harrier landing strip to airlift "packages" of 24 Airmobile Brigade in "cabs" from one location to another.' (Author)

Pumas use the West Tofts Harrier landing strip to airlift "packages" of 24 Airmobile Brigade in "cabs" from one location to another at night. Sitting on the jump seat of a Chinook immediately under the front rotor blades at start-engines time is akin to bouncing up and down in a food mixer. Underneath the helicopter the downdraught from the two gigantic egg whisks is gale force. The 60 ft-diameter rotor blades attached to the two 3,750 shp Lycoming T-55-L-11 turboshafts can chop a man in half and toss him 200 m. They scatter loose grass like confetti. The 18 Squadron crew from Laarbruch lifts off and heads for the first RV [radar vector]. In the rear fuselage the two loadmasters await their visitors from deep in the forests, but the pilots are kept "handbagging around" while those on the ground sort themselves out. Out of the blackness a white "T" finally appears. A quick glance into the nocturnal world with the NVGs identified rows of Land Rovers and trailers parked to the rear of the lights. Though it is the dead of night, it is like looking through a green sweet wrapper in broad daylight. Blobs of white right and left are revealed as Pumas and Chinooks on identical missions. In the distance the red street lights of Thetford turn massively white, standing out like a futuristic city. Electromagnetic particles dance along the forward rotor blades of the Chinook like St Elmo's Fire.

'Freezing gusts of air whip through the Chinook as the rear ramp is lowered. Down first to pick up the troops, a short hop to attach the underslung vehicles, and up and away to the next RV with a motorbike and 22 Milan-armed soldiers aboard. Dangling below the cargo hole, a Land Rover and trailer hanging on umbilical chords swing gently like acrobats on a trapeze. Off to the right a Puma mirrors our every move. However, one side of the "T" lights has been

placed only 15 m from the trees instead of the regulation 30 m, and it is only after some discussion that the Chinook crew decide to continue. The deciding factors are the NVGs and the crew's familiarity with the location, gained during the daylight hours.

'All through the night the white and green Chinooks, and Pumas, will fly back and forth, until, finally, the entire brigade will have been exfiltrated from their former locations and deposited into new areas. It will enable the next phase of the exercise to begin. At first light the unmistakable sound of Lycoming turboshafts and Turbomeca Turmos will be heard champing at

'Puma crews were practising the whole gamut of NBC drill, carrying their PV (portable ventilators) to and from their helicopters . . .' (Author)

the bit once more. The exercise reaches a new phase, and 24 Airmobile Brigade will again enter the eye of the storm brewing in STANTA's forest-green grotto.

(A week after this exercise, 24 Airmobile Brigade was put on standby for deployment to Bosnia.)

THE GRAND-CHILDREN

What do they know of the siren's wail
In the pitch-dark gloom of a blacked-out town,
The droning waves of the enemy planes
And the crump as a bomb comes whistling down,
The angry bark of the Ack Ack guns
And the shrapnel tinkling round one's feet,
The acrid smell and the orange glare
As the flames consume a burning street,
And the blackened shells of a city's homes,
Still smouldering in the dawn's cold light,
While the citizens calmly carry on
And brace themselves for another night?

What do they know of the waiting hours
On the grass outside the dispersal huts,
The race to the 'planes when the scramble starts
And the dried-up mouths and the twisting guts?
A Hurricane swoops and fires its guns,
A short sharp burst of chattering sound,
The sickening dive and the screaming power
As a Messerschmitt spirals to the ground,
The sky a mêlee of wings and guns
As the aircraft weave in the furious fray,
And the pilot's thought when the the fight is done
— How many friends has he lost today?

What do they know of the briefing room
And the questioning eyes of the bomber crews
As they learn of the target for tonight
And scorn to flinch as they hear the news
That they have to go though the murderous flak
Of a well-defended German town,
And they know they must fly to Hell and back
To drop their dreadful cargo down,
The quickening pulse as the engines roar
And their journey starts in the fading light,
And the courage of men who know the odds
Against surviving this lethal flight.

What do they know of the pain of loss
As the Grim Reaper takes his pay,
The lives so young that were his reward
As the price of the freedom we have today.
We who are left have memories bright
Of those who shone in our darkest hours,
But we pray, dear lord, that our little ones
Will never have memories such as ours.

Audrey Grealy

LINGUA FRANCA
(Lore of the Service)

ack-ack	flak, anti-aircraft fire
Air Cdre	Air Commodore
ACM	Air Chief Marshal
ADGB	Air Defence of Great Britain
AM	Air Marshal
ASV	air-to-surface-vessel radar
AVM	Air Vice-Marshal
'Banana jet'	Buccaneer aircraft (because of its shape)
batman	From the French *bat*, meaning pack saddle. A male or female mess steward responsible for an officer's well-being on base
BFTS	British Flying Training School
bind, a	a tiresome nuisance
Big City	Berlin
blanket stackers	non-flying personnel
Blitz, a solid	a large formation of enemy aircraft
blunties, lump of	non-flying personnel (not sharp)
boomerang	abort, turn back
brassed off	fed up
brevet	flying badge
Cab Rank	Small formations of fighter-bombers on immediate standby for close tactical support
Cheshire	Relief operation in Bosnia
chop, get the	be shot down/killed
chopper	helicopter
cheesed off	very fed up
cookie	4,000 lb bomb
Corporate	RAF operations in the Falklands Conflict
Crossbow	Offensive and defensive measures against the V1 flying bomb
cushy trip	easy mission
Deny Flight	Joint USAF/RAF and Allied air forces operation to prevent air activity in Bosnia
DFC	Distinguished Flying Cross
Dicey	dangerous
ditch, to	put down on water (in the 'drink')
Diver	codename for V1 flying bomb
drink, the	the sea
DSO	Distinguished Service Order
duff gen	bad information
eggs	bombs
Erk	beginner
Few, The	All the RAF pilots in the Battle of Britain
Fg Off	Flying Officer
FIDO	Fog investigation and dispersal operation
flak	German term for anti-aircraft fire
flap	panic
Flt Sgt	Flight Sergeant
Flying Tadpole or Flying Suitcase	Handley Page Hampden, so called because its fuselage was only 3 feet wide at its widest point, and its tail surfaces were carried on a long, thin boom
Gardening	minelaying
Gone for a Burton	Shot down or expired. This expression was derived from a Burton Ales advertisement which always showed a group of people with one person obviously missing – the captain had gone for a Burton
gong	medal
goolie chit	Piece of paper bearing HM Government promise to pay the bearer a sum of money (£5,000 during the Gulf War) providing the airman is returned unharmed, with his 'goolies' or testicles, still attached. Dates from when the RAF policed the Empire and Middle East between the wars

Granby	RAF operations in the Gulf War
gremlin	A mythical mischievous creature invented by the RAF, to whom is attributed the blame for anything that goes wrong in the air or on the ground. There are different sorts of gremlins skilled in different sorts and grades of evil. The origin of the term is obscure, but has been stated variously to go back as far as the Royal Naval Air Service; to have some connection with the RAF in Russia and the Kremlin; and to have come from India, where, it is alleged, in the early 1920s an officer was opening a bottle of Fremlin's Ale when the overheated gas blew out the cork, taking him by surprise. Meaning to say: 'A goblin has jumped out of my Fremlin's', he spoonerised and said: 'A gremlin has jumped out of my Foblin's'. In his book *It's a Piece of Cake!: RAF Slang Made Easy* (Sylvan Press, c 1942), Sqn Ldr C.H. Ward-Jackson adds: 'Officers and airmen who are on the right side of the gremlins are thought very highly of by Station Commanders but are objects of suspicion among their fellows'.
Hallybag	Handley Page Halifax four-engined bomber
Happy Valley	The Ruhr
HAS	hardened aircraft shelter
Heath Robinson, W.	Artist and illustrator (1872–1944) famous for his comical drawings of ingenious makeshift mechanical contrivances or structures. His name is used as an adjective to describe weird or imaginary devices
hit the silk	bale out using a parachute
IFF	identification friend or foe
kite	aeroplane
Kite, Fg Off	archetypal flamboyant flying type
LAC	leading aircraftman
linies	non-flying personnel
LMF	lack of moral fibre
Mae West	lifejacket, named after the well-endowed American actress
Mickey Mouse	bomb-aiming equipment
Manna	Air supply mission to Holland, April/May 1945
Market Garden	Airborne operations, Arnhem, September 1944
mess	Possibly from the latin *mensa* (table) or Old French *mes* (dish of food)
met gen	meteorological information
milk run	regular run of operations to a particular target (US, easy mission)
Millenium	One of three 1,000-bomber raids on German cities, May-June 1942
mud mover	Tornado GR.1
mufti	civilian clothes, Indian word
Newhaven	flares dropped by PFF
Nickels	propaganda leaflets
Noball	V2 rocket and V1 flying bomb sites
Ops	operations
other ranks	ranks other than commissioned officers
OTU	Operational Training Unit
Paramatta	flares dropped by PFF
plumbers	armourers
Plt Off	Pilot Officer
prang	crash
press the tit	press the bomb release (bomber) or firing button (fighter)
Prune, Plt Off	A fictitious character who was everything a pilot should not be
RCM	radio countermeasures
Rock Apes	members of the RAF Regiment
SAM	surface-to-air missile
scopies	fighter controllers
scramble	take-off at once
scribblies	non-flying personnel (not sharp)
snowdrop	RAF military policeman
sortie	operational flight by a single aircraft
sprog	inexperienced person
WAAF	Women's Auxiliary Air Force (member of)
Wanganui	Skymarking a target using flares dropped blindly using H_2S
water rats	firemen
Wg Cdr	Wing Commander, known familiarly as 'Wingco'
Wimpy	Vickers Armstrongs Wellington twin engined bomber (from J. Wellington Wimpy – character in 'Popeye'.)
Window	Thin metallic strips dropped by Bomber Command to disrupt enemy radar screens
wizard	first-class
wizard prang	big crash
WOP/AG	wireless-operator/air gunner

BIBLIOGRAPHY

Airmen's Song Book Edited by C.H. Ward-Jackson & Leighton Lucas, Blackwood & Sons, 1967.
Behind the Hangar Doors Philip Congdon, Sonik Books, 1985
The Falcon, The Magazine of No 4 BFTS, compiled by Capt Bill McCash
Gulf Air War Debrief Aerospace Publishing, 1991
Handle with Care R. Anderson & D. Westmacott, 1946
Into Battle With 57 Sqn Roland A. Hammersley DFM, privately published, 1992
Kites & Kriegies Basil S. Craske RAFVR, privately published
Marker, The Pathfinder Association Magazine
Pathfinder AVM D.C.T. Bennett CB, CBE, DSO, Panther, 1960
Piece of Cake Geoff Taylor, George Mann, 1956
Punch
Royal Air Force RAF Public Relations Magazine, Nos 4, 5, 6, & 7, Editor, Sqn Ldr Jake Mclaughlin MBE BEM
The Royal Air Force: An illustrated History Michael Armitage, Arms & Armour 1993
The RAF in Action: From Flanders to the Falklands Robert Jackson, Blandford Press, 1985
Take-Off magazine Aerospace Publishing, 1993
Thunder & Lightning – The RAF in the Gulf – Personal Experiences of War Charles Allen, HMSO, 1991
Tornado Down John Peters and John Nichol, Michael Joseph, 1992
Women in RAF Blue Sqn Ldr Beryl E. Escott, Patrick Stephens, 1989